Revise GCSE

Biology

Ian Honeysett

Contents

This book and

	AQA	Edexcel
Web Address	www.aqa.org.uk	www.edexcel.com
Specification Number	4401	2BI01
Exam Assessed Units and Modules At least 40% of assessment must be carried out at the end of the course. For students starting the GCSE course from September 2012 onwards, all assessment (100%) must take place at the end of the course.	**Unit 1: Biology 1** 1 hr 60 marks 25% of GCSE **Unit 2: Biology 2** 1 hr 60 marks 25% of GCSE **Unit 3: Biology 3** 1 hr 60 marks 25% of GCSE All papers feature structured and closed questions.	**Unit B1** 1 hr 60 marks 25% of GCSE **Unit B2** 1 hr 60 marks 25% of GCSE **Unit B3** 1 hr 60 marks 25% of GCSE All papers feature objective, short answer and extended writing questions.
Controlled Assessment Covering: · Research, planning and risk assessment · Data collection · Processing, analysis and evaluation	**Unit 4: Controlled** 1hr 35 min, plus time for research / data collection 50 marks 25% of GCSE	**Unit PCA** 3 hrs, plus preparation time 50 marks 25% of GCSE
Chapter Map*		
1 **Organisms in action**	B1.1, B1.2, B3.3	B1, B3
2 **Health and disease**	B1.1, B1.3	B1, B3
3 **Genetics and evolution**	B1.7, B1.8, B2.7, B2.8	B1, B2
4 **Organisms and environment**	B1.4, B1.5, B1.6, B2.8, B3.4	B1, B3
5 **Cells and molecules**	B2.1, B2.2, B2.5, B2.6, B2.7, B3.1	B2
6 **Sampling organisms**	B2.3, B2.4, B3.4	B2
7 **Physiology**	B2.2, B2.5, B3.1, B3.2	B2, B3
8 **Use, damage, repair**	B3.1, B3.3	B1, B3
9 **Animal behaviour**		B3
10 **Microbes**	B1.7, B2.7, B3.4	B2, B3

* There are tick charts throughout the book to show which particular sub-topics in each chapter are relevant to your course.

your GCSE course

OCR A	OCR B	WJEC	CCEA
www.ocr.org.uk	www.ocr.org.uk	www.wjec.co.uk	www.ccea.org.uk
J243	J263	600/0895/7	G09
Modules B1, B2 and B3 1 hr 60 marks 25% of GCSE **Modules B4, B5 and B6** 1 hr 60 marks 25% of GCSE **Module B7** 1 hr 60 marks 25% of GCSE All papers feature objective style and free response questions.	**Modules B1, B2 and B3** 1hr 15 min 75 marks 35% of GCSE **Modules B4, B5 and B6** 1hr 30 min 85 marks 40% of GCSE Includes a data response section worth 10 marks, which assesses AO3 All papers feature structured questions.	**Biology 1** 1 hr 60 marks 25% of GCSE **Biology 2** 1 hr 60 marks 25% of GCSE **Biology 3** 1 hr 60 marks 25% of GCSE All papers feature structured questions involving some external prose.	**Unit B1** Higher: 1 hr 30 min 100 marks Foundation: 1 hr 15 min 80 marks 35% of GCSE **Unit B2** Higher: 1 hr 45 min 115 marks Foundation: 1 hr 30 min 90 marks 40% of GCSE All papers feature structured questions.
Unit A164 Approx. 4.5–6hrs 64 marks 25% of GCSE	**Unit B733** Approx. 7 hrs 48 marks 25% of GCSE	**Unit CA** Approx. 7.5 hrs, plus time for initial research 48 marks 25% of GCSE	**Unit 3** 1 hr, plus time for planning, risk assessment and data collection 45 marks 25% of GCSE
B2, B5, B6, B7	B1, B2, B5	B1, B3	B1, B2
B2, B7	B1, B6	B1, B2, B3	B1, B2
B1, B3	B1, B2, B3, B6	B1, B2	B2
B3, B7	B2, B4, B6	B1, B2	B1
B1, B4, B5, B7	B3, B4, B6	B1, B2, B3	B1, B2
B4	B4	B1, B2, B3	B1
B2, B7	B3, B4, B5, B6	B2, B3	B1, B2
B7	B5	B3	B1, B2
B6			
B4, B7	B6	B1, B3	B2

Preparing for the exams

What will be assessed

In your science exams and controlled assessment you are assessed on three main criteria called assessment objectives:

- Assessment Objective 1 (AO1) – tests your ability to **recall**, select and communicate your knowledge and understanding of biology.
- Assessment Objective 2 (AO2) – tests your ability to **apply** your skills, knowledge and understanding of biology in practical and other contexts.
- Assessment Objective 3 (AO3) – tests your ability to **analyse** and **evaluate** evidence, make reasoned judgements and draw conclusions based on evidence.

The exam papers have a lot of AO1 and AO2 questions and some AO3 questions. The controlled assessments focus mainly on AO2 and AO3.

To do well on the exams, it is not enough just to be able to recall facts. You must be able to apply your knowledge to different scenarios, analyse and evaluate evidence and formulate your own ideas and conclusions.

Planning your study

The 40% terminal rule for GCSE means that 40% of your assessment must be taken at the end of the course. However, this means that 60% can be taken before the end – so it is important to have an organised approach to study and revision throughout the course.

- After completing a topic in school or college, go through the topic again using this guide. Copy out the main points on a piece of paper or use a pen to highlight them.
- Much of memory is visual. Make sure your notes are laid out in a logical way using colour, charts, diagrams and symbols to present information in a visual way. If your notes are easy to read and attractive to the eye, they will be easier to remember.

- A couple of days later, try writing out the key points from memory. Check differences between what you wrote originally and what you wrote later.
- If you have written your notes on a piece of paper, make sure you keep them for revision later.
- Try some questions in the book and check your answers.
- Decide whether you have fully mastered the topic and write down any weaknesses you think you have.

How this book will help you

This complete study and revision guide will help you because:

- it contains the essential content for your GCSE course without the extra material that will not be examined
- there are regular, short progress checks so that you can test your understanding
- it contains sample GCSE questions with model answers and notes, so that you can see what the examiner is looking for.
- it contains exam practice questions so that you can confirm your understanding and practise answering exam-style questions
- the summary table on pages 3–4, and the exam-board signposting throughout the book, will ensure that you only study and revise topics that are relevant to your course.

Six ways to improve your grade

1. Read the question carefully

Many students fail to answer the actual question set. Perhaps they misread the question or answer a similar question that they have seen before. Read the question once right through and then again more slowly. Underline key words in the question as you read through it. Questions at GCSE often contain a lot of information. You should be concerned if you are not using the information in your answer.

Take notice of the command words used in questions and make sure you answer appropriately:

- **State:** A concise, factual answer with no description or explanation
- **Describe:** A detailed answer that demonstrates knowledge of the facts about the topic
- **Explain:** A more detailed answer than a description; give reasons and use connectives like 'because'
- **Calculate:** Give a numerical answer, including working and correct units
- **Suggest:** A personal response supported by facts

2. Give enough detail

If a part of a question is worth three marks, you should make at least three separate points. Be careful that you do not make the same point three times, but worded in a slightly different way. Draw diagrams with a ruler and label with straight lines.

3. Be specific

Avoid using the word 'it' in your answers. Writing out in full what you are referring to will ensure the examiner knows what you are talking about. This is especially important in questions where you have to compare two or more things.

4. Use scientific language correctly

Try to use the correct scientific language in your answers. The way scientific language is used is often the difference between successful and unsuccessful answers. As you revise, make a list of scientific terms you come across and check that you understand what they mean. Learn all the definitions. These are easy marks and they reward effort and good preparation.

5. Show your working

All science papers include calculations. Learn a set method for solving a calculation and use that method. You should always show your working in full. That way, if you make an arithmetical mistake, you may still receive marks for applying the correct science. Check your answer is given to the correct level of accuracy (significant figures or decimal places) and give the correct units.

6. Brush up on your writing skills

Your exam papers will include specific questions for which the answers will be marked on both scientific accuracy and the quality of the written communication. These questions are worth 6 marks, but it does not matter how good the science is, your answer will not gain full marks unless:

- the text is legible and the spelling, punctuation and grammar are accurate so that your meaning is clear
- you have used a form and style of writing that is fit for purpose and appropriate to the subject matter
- you have organised information in a clear and logical way, correctly using scientific vocabulary where appropriate.

These questions will be clearly indicated on the exam papers.

> Exam papers are scanned and marked on a computer screen. Do not write outside the answer spaces allowed, or your work may not be seen by the examiner. Ask for extra paper if you need it. Choose a black pen that will show up – one that photocopies well is a good choice.

How Science Works

The science GCSE courses are designed to help develop your knowledge of certain factual details, but also your understanding of 'How Science Works'.

'How Science Works' is essentially a set of key concepts that are relevant to all areas of science. It is concerned with four main areas:

Data, evidence, theories and explanations

- science as an evidence-based discipline
- the collaborative nature of science as a discipline and the way new scientific knowledge is validated
- how scientific understanding and theories develop
- the limitations of science
- how and why decisions about science and technology are made
- the use of modelling, including mathematical modelling, to explain aspects of science

Practical skills

- developing hypotheses
- planning practical ways to test hypotheses
- the importance of working accurately and safely
- identifying hazards and assessing risks
- collecting, processing, analysing and interpreting primary and secondary data
- reviewing methodology to assess fitness for purpose
- reviewing hypotheses in light of outcomes

Communication skills

- communicating scientific information using scientific, technical and mathematical language, conventions, and symbols.
- using models to explain systems, processes and abstract ideas

Applications and implications of science

- the ethical implications of biology and its applications
- risk factors and risk assessment in the context of potential benefit

You will be taught about 'How Science Works' throughout the course in combination with the scientific content. Likewise, the different exam boards have included material about 'How Science Works' in different parts of their assessment.

'How Science Works' will be assessed in the controlled assessment, but you will also get questions that relate to it in the exams. If you come across questions about unfamiliar situations in the exam, do not panic and think that you have not learnt the work. Most of these questions are designed to test your skills and understanding of 'How Science Works', not your memory. The examiners want you to demonstrate what you know, understand and can do.

1 Organisms in action

The following topics are covered in this chapter:

- A balanced diet
- Homeostasis 1
- Homeostasis 2
- Hormones and reproduction
- Responding to the environment
- The nervous system
- Plant responses

1.1 A balanced diet

After studying this section you should be able to:

- identify the components of a balanced diet and explain their roles
- realise that dietary requirements vary in different people
- calculate the EAR for protein in the diet and appreciate that there are different types of proteins
- explain how different food substances are stored.

Different food molecules

AQA	B1	✓
OCR B	B1	✓
CCEA	B1	✓

All organisms need food to survive. Food provides raw materials for growth and energy.

We take in our food ready-made as complicated organic molecules. These food molecules can be placed into seven main groups (see table below).

> **KEY POINT**
>
> A balanced diet needs the correct amounts of each of the types of food molecules in the table.

Remember, a balanced diet is the 'correct' amount of each food type. In exams, candidates can lose a mark by saying a balanced diet contains 'enough' food.

You can use this rhyme to remember the seven types of food molecules: when my parents cook, vegetables feel funny.

This stands for: water, minerals, proteins, carbohydrates, vitamins, fats and fibre.

Food type	Made up of	Use in the body
Water	Hydrogen and oxygen	Prevents dehydration
Minerals	Different elements, e.g. iron	Iron is used to make haemoglobin
Proteins	Long chains of amino acids	Growth and repair
Carbohydrates	Simple sugars, e.g. glucose	Supply or store of energy
Vitamins	Different structures, e.g. vitamin C	Vitamin C prevents scurvy
Fats	Fatty acids and glycerol	Rich store of energy
Fibre	Cellulose	Prevents constipation

Different people may have slightly different diets due to a number of factors, including:

- **age** and **gender**
- level of **activity**
- personal preference, e.g. whether they are **vegetarians** or **vegans**
- medical reasons, e.g. if they have any food **allergies**.

Protein in the diet

OCR B   B1   ✓

Proteins are needed for growth and so it is important to eat the correct amount. This is called the **estimated average requirement (EAR)** and can be calculated using the formula:

> **KEY POINT**
>
> **EAR (g) = 0.6 × body mass (kg)**

Too little protein in the diet causes the condition called **kwashiorkor**. This is common in developing countries due to overpopulation and lack of money to improve agriculture.

The EAR is an estimate of the mass of protein needed per day based on the average person. The EAR for protein might be affected by factors such as age, pregnancy or breast feeding (lactation).

To get an A* you must be able to analyse data about different types of protein.

Although proteins cannot be stored in the body, some amino acids can be converted by the body into other amino acids. However, there are some amino acids that can only be obtained from the diet – they are called **essential amino acids**.

Proteins from meat and fish are called **first class proteins**. They contain all the essential amino acids that cannot be made by the body.

Plant proteins are called **second class proteins** as they do not contain all the essential amino acids.

The amino acid that is in shortest supply is called the limiting amino acid. This will restrict the growth of the person.

Storing food

OCR B   B1   ✓
WJEC   B1   ✓

If you eat too much fat and carbohydrate, the excess is stored in the body.

> **KEY POINT**
>
> Carbohydrates are stored in the **liver** as **glycogen** or are converted into fats.

There is a limit to how much glycogen the liver can store, but fat storage is not so limited.

Fats are stored under the skin and around organs as **adipose tissue**.

Although proteins are essential for growth and repair, they cannot be stored in the body.

1. What are proteins made of?
2. Tom has a mass of 55 kg. What is his EAR for protein?
3. What is kwashiorkor?
4. Why is it important to have enough fibre in the diet?
5. Look at the table showing details of four foods.

Food	Protein quality rating	Limiting amino acid
Egg	0.98	None
Beef	0.77	None
Wheat	0.62	Lysine
Peas	0.49	Methionine

a) Why do egg and beef have a higher protein quality rating?

b) Why do many diets often involve eating peas with wheat?

1. Amino acids
2. $0.6 \times 55 = 33$ g
3. Kwashiorkor is a deficiency disease caused by a lack of protein in the diet.
4. To move food along the gut (avoid constipation).
5. a) Egg and beef are first class proteins so contain all the essential amino acids.
   b) Peas and wheat are both deficient in a different amino acid so they complement each other.

1.2 Homeostasis 1

LEARNING SUMMARY

After studying this section you should be able to:

- explain the importance of homeostasis in the body
- explain what is meant by the terms hormone and negative feedback
- explain how blood sugar levels are controlled
- describe the causes and symptoms of diabetes.

Principles of homeostasis

AQA	B1	✓
OCR A	B2	✓
OCR B	B1	✓
EDEXCEL	B1	✓
WJEC	B1	✓

It is vital that the internal environment of the body is kept constant. This state is called **homeostasis**.

The different factors in the body that need to be kept constant include:

- water content
- temperature
- sugar content
- mineral (ion) content.

Many of the mechanisms that are used for homeostasis involve **hormones**.

KEY POINT

Hormones are chemical messengers that are carried in the blood stream.

They are released by glands and passed to their target organ.

Hormones take longer to have an effect compared to nerves, but their responses usually last longer.

Negative feedback

OCR A	B2	✓
OCR B	B1	✓
EDEXCEL	B1	✓
WJEC	B1	✓

The homeostasis control mechanisms in the body work by **negative feedback**. This means that **receptors** in the body detect a change in the body. These changes are then **processed** in the body. Then **effectors** bring about a response that reverses the change so that the normal level is restored.

This is much like many artificial control systems such as the temperature control in a house.

Many control devices in the home work by negative feedback. Thinking about how they work might help you to remember what is meant by negative feedback.

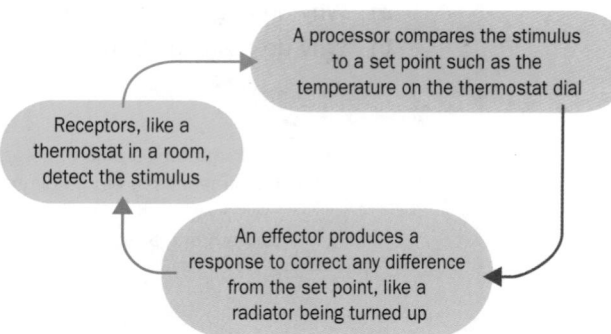

Receptors, like a thermostat in a room, detect the stimulus

A processor compares the stimulus to a set point such as the temperature on the thermostat dial

An effector produces a response to correct any difference from the set point, like a radiator being turned up

Controlling blood sugar levels

AQA	B3	✓
OCR A	B7	✓
OCR B	B1	✓
EDEXCEL	B1	✓
WJEC	B1	✓
CCEA	B1	✓

It is vital that the sugar or glucose level of the blood is kept constant:

- If it gets too low then cells will not have enough glucose to use for respiration.
- If it is too high then glucose may start to be excreted in the urine.

> **KEY POINT**
>
> **Insulin** is the hormone that controls the level of glucose in the blood.

Insulin is made in the pancreas. When glucose levels are too high, more insulin is released. The insulin acts on the **liver** causing it to convert excess glucose into **glycogen** for storage.

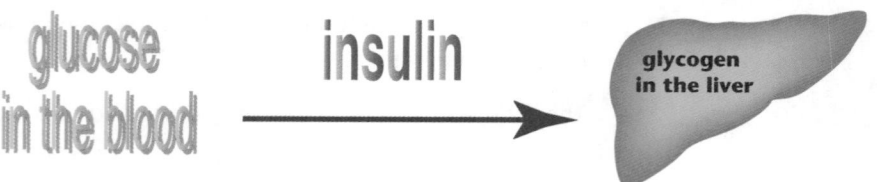

glucose in the blood → insulin → glycogen in the liver

The role of glucagon

AQA	B3	✓
EDEXCEL	B1	✓
CCEA	B1	✓

Some systems have two or more effectors, which can work in opposite directions. This means that the response can happen much faster.

> Try not to get confused between glucose, glucagon and glycogen. They have similar names and you must be careful that the examiner knows which you mean!

If blood sugar levels are too high then the hormone insulin is released. If levels drop below normal, the pancreas releases another hormone called glucagon.

Glucagon will cause the glycogen that is stored in the liver to be converted back to glucose.

Diabetes

AQA	B3	✓
OCR A	B7	✓
OCR B	B1	✓
EDEXCEL	B1	✓
WJEC	B1	✓
CCEA	B1	✓

> **KEY POINT**
>
> People who cannot control their blood sugar levels have a condition called diabetes.

> To get an A*, you must be able to look at graphs of blood sugar levels and work out when a person with diabetes did exercise, ate a meal or injected themselves with insulin.

This often causes blood sugar levels to be too high and so glucose is excreted from the body in the urine. This can be tested for using testing strips that are dipped into the urine and change colour if glucose is present.

There are two types of diabetes:

- **Type 1** – A genetic disease that is caused by the pancreas failing to make enough insulin. It is treated with regular insulin injections in order to control the level of glucose in the blood. People with Type 1 diabetes also need to control their diet carefully.
- **Type 2** – Caused by the cells in the body failing to respond to insulin. This is controlled by making sure that the person does not eat too much carbohydrate in meals.

People with Type 1 diabetes need to inject themselves with insulin. It is important that they get the dose right so they have to test their blood to see how much sugar is present. This will vary depending on:

- when they last had a meal
- how much exercise they have done recently.

> **PROGRESS CHECK**
>
> 1. What is a hormone?
> 2. Where in the body is insulin made and released?
> 3. What effect does insulin have on the liver?
> 4. Write down one way that a person can be tested for diabetes.
> 5. What is the difference in function between glucose, glycogen and glucagon?
> 6. Suggest what effect exercise is likely to have on the blood sugar level.
>
> 6. Exercise will reduce the blood sugar level as glucose is used up in respiration.
> 5. Glucose is a sugar used for respiration in the body, glycogen is a storage carbohydrate and glucagon is a hormone that converts glycogen into glucose.
> 4. One test for diabetes is using urine testing sticks to check for glucose.
> 3. Insulin causes the liver to convert glucose into the storage carbohydrate glycogen.
> 2. Insulin is made and released in the pancreas.
> 1. A hormone is a chemical messenger that causes a response in the body.

1.3 Homeostasis 2

LEARNING SUMMARY	After studying this section you should be able to:
	• explain why and how body temperature is regulated
	• describe how the water content of the body is regulated.

Temperature regulation

AQA	B3	✓
OCR A	B7	✓
OCR B	B1	✓
EDEXCEL	B1	✓
WJEC	B1	✓

> **KEY POINT**
>
> It is important to keep our body temperature at about **37°C**.

If the body temperature gets too low this is called **hypothermia** and this can be fatal. If the blood temperature gets too high it could lead to **heat stroke** and **dehydration**.

The body temperature is monitored by the brain and if it varies from 37°C, various changes are brought about.

When we feel hot we need to lose heat faster, as our core body temperature is in danger of rising.

We do this by:

- **Sweating** – as water evaporates from our skin, it absorbs heat energy from the body/skin. This cools the skin and the body loses heat.
- Sending more blood to the skin, so that more heat is lost by radiation. This causes the skin to look red.

When we feel too cold we are in danger of losing heat too quickly and cooling down. This means we need to conserve our heat to maintain a constant 37°C.

We do this by:

- **Shivering** – rapid contraction and relaxation of body muscles. This increases the rate of respiration and more energy is released as heat.
- Sending less blood to the skin, so the blood is diverted to deeper within the body to conserve heat and because of this the skin looks pale.
- Sweating less.

Body temperature needs to be 37°C because it is the best temperature for **enzymes** in the body, which control the rate of chemical reactions in the body, to work.

Any change in body temperature is detected by the **thermoregulatory centre** in the brain. This will bring about the correction mechanisms.

> In addition to knowing about temperature regulation mechanisms, WJEC candidates must have a detailed knowledge of skin structure, including the epidermis, dermis and subcutaneous tissue.

Make sure you describe vasodilation and vasoconstriction as the widening or narrowing of blood vessels. Many students lose marks by saying that blood vessels move towards or away from the skin!

Body temperature is detected in the thermoregulatory centre of the brain

If the temperature rises various changes occur.

Increased sweat production which evaporates and blood vessels in the skin widen allowing blood vessels nearer the skin surface. **(vasodilation)**

If the temperature drops changes occur to slow down heat loss.

Decreased sweat production the blood is diverted to deeper blood vessels in the skin and close down. **(vasoconstriction)**

Body temperature returns to normal.

Control of water balance

OCR A	B2	✓
OCR B	B1	✓
EDEXCEL	B3	✓

It is important to control the amount of water in the body otherwise the blood can become too concentrated or diluted. This is done by making sure that over a certain period of time we take in the same amount of water that we give out.

The body gains water by:

- drinking
- eating food
- respiration which releases water.

The body loses water through:

- sweating
- breathing
- faeces
- excreting urine.

Most of the regulation of water content is done by the **kidneys** altering the volume and concentration of the urine. The kidneys control the water balance of the body by filtering the blood to remove all small molecules.

Then useful molecules (e.g. glucose), and a certain amount of water and salts, are taken back into the blood to keep their levels in balance. The remaining waste is stored in the bladder as urine.

To get an A*, you must make sure that you remember how ADH works. More ADH makes the kidney reabsorb more water and so make less urine. Many students get this confused.

> ### KEY POINT
>
> The amount of water that is taken back into the blood is controlled by a hormone called **antidiuretic hormone (ADH)** which is released by the **pituitary gland**.

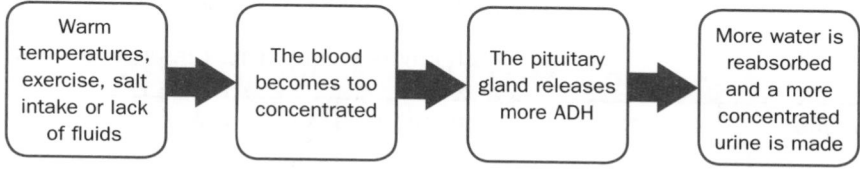

Warm temperatures, exercise, salt intake or lack of fluids → The blood becomes too concentrated → The pituitary gland releases more ADH → More water is reabsorbed and a more concentrated urine is made

Different drugs can alter ADH release:

- **Alcohol** reduces ADH release and can cause too much urine to be made.
- **Ecstasy** can cause the opposite effect.

1.4 Hormones and reproduction

LEARNING SUMMARY

After studying this section you should be able to:

- describe the role of hormones in controlling reproduction
- describe how the menstrual cycle is regulated
- describe how hormones can be used to manipulate fertility.

Reproductive hormones

AQA	B1	✓
OCR B	B5	✓
EDEXCEL	B3	✓
CCEA	B2	✓

Hormones are responsible for controlling many parts of the reproduction process.

This includes:

- the development of the sex organs
- the production of sex cells
- controlling pregnancy and birth.

The main hormones controlling these processes are shown in the table.

Hormone	Male or female	Produced by	Main function
Testosterone	Male	Testes	Stimulates the male secondary sexual characteristics.
Oestrogen	Female	Ovaries	Stimulates the female secondary sexual characteristics. Repairs the wall of the uterus.
Progesterone	Female	Ovaries and placenta	Prevents the wall of the uterus breaking down.

Testosterone and oestrogen control the changes occurring in the male and female bodies at puberty. These changes are the secondary sexual characteristics.

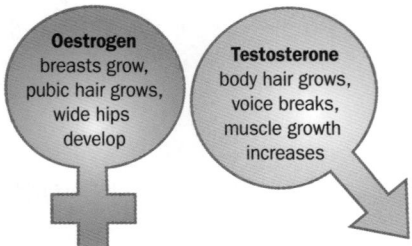

The secondary sexual characteristics also include the production of the sex cells. In the male they are sperm and in females they are eggs.

After puberty in the male, sperm are produced continuously, but in the female one egg is usually released about once a month.

This means that oestrogen and progesterone levels vary at different times in the monthly or menstrual cycle.

- Oestrogen levels are high in the first half of the cycle. The oestrogen prepares the wall of the uterus to receive a fertilised egg. It does this by making it thicker and increasing its supply of blood. It also triggers the release of an egg. This is called ovulation.
- Progesterone is high in the second half of the cycle. It further repairs the wall of the uterus and stops it breaking down.

Changes occur during the monthly cycle.

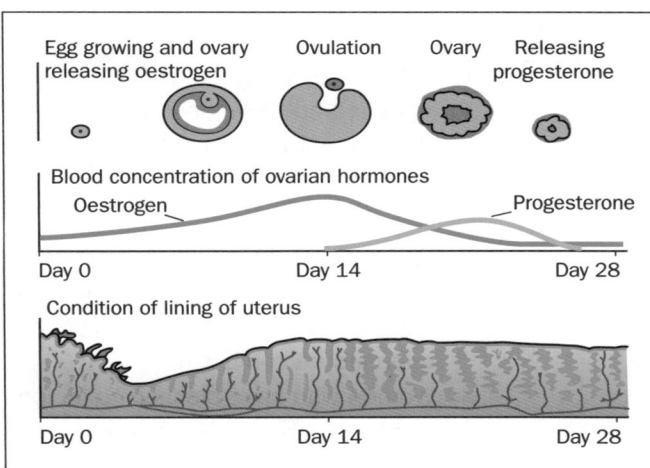

If a question gives a diagram of the menstrual cycle and asks you when ovulation occurs do not automatically say day 14. Look for the peak of the oestrogen curve or the increase in progesterone. All women are different!

The production of oestrogen and progesterone is controlled by the release of other hormones. These hormones are called luteinising hormone (LH) and follicle stimulating hormone (FSH) and both are made in the pituitary gland in the brain.

Control of reproduction.

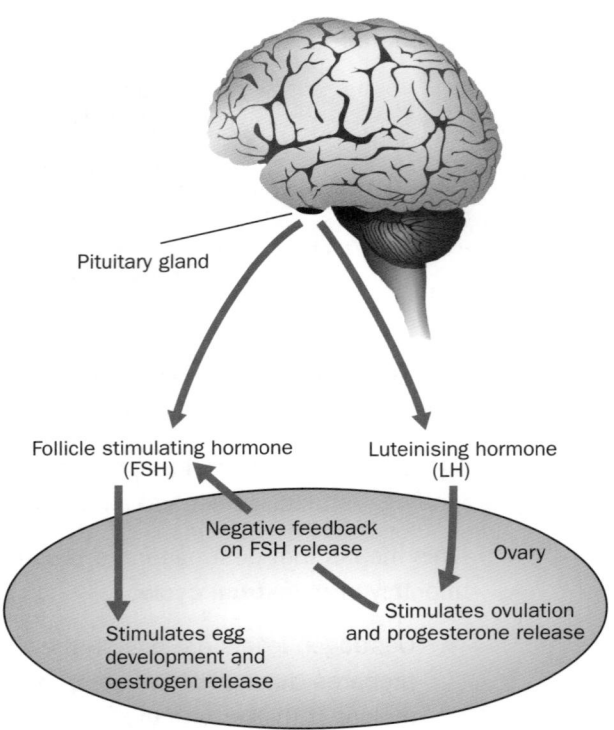

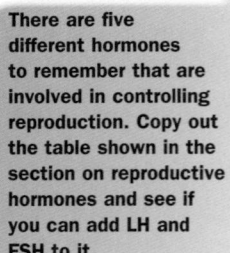

There are five different hormones to remember that are involved in controlling reproduction. Copy out the table shown in the section on reproductive hormones and see if you can add LH and FSH to it.

Treating infertility

AQA	B1	✓
OCR B	B5	✓
EDEXCEL	B3	✓
CCEA	B2	✓

About one in seven of all couples have difficulty having a baby. There are many reasons for infertility and these include:

- a blockage in the Fallopian tubes or in the sperm ducts
- eggs are not developed or released from ovaries
- there is not enough fertile sperm produced by testes.

It is now possible to treat some of these cases of infertility by using hormones:

- Women who do not develop or release eggs from their ovaries can take a fertility drug. This contains hormones that are similar to FSH. The drugs stimulate the production of eggs and sometimes a number of eggs are released each month.
- Women who have blocked Fallopian tubes can be treated with fertility drugs and a number of eggs are removed from their body. The eggs can then be fertilised by sperm outside the body. The embryo can then be put back inside the uterus. This process is called in vitro fertilisation (IVF).

Decreasing fertility

AQA	B1	✓
OCR B	B5	✓
EDEXCEL	B3	✓
CCEA	B2	✓

Some women may want to stop themselves becoming pregnant.
They take drugs that are called oral contraceptives.

These drugs contain different amounts of the hormones oestrogen and progesterone.

CCEA candidates also need to know how non-hormonal methods of contraception work and some of the advantages and disadvantages of each method.

They prevent the pituitary gland releasing FSH. This means that the ovary will not produce eggs.

PROGRESS CHECK

1. Where is testosterone made and what does it do?
2. What is ovulation?
3. What is IVF?
4. How do oral contraceptives work?
5. Normally the progesterone level falls towards the end of the menstrual cycle. Why is it important that it stays high if the egg has been fertilised?
6. The 'morning after pill' contains very high levels of oestrogen. Suggest why many people think that it should not be used regularly for contraception.

1. Testosterone is made in the testes and it causes the development of secondary sexual characteristics in the man including the production of sperm.
2. Ovulation is the release of an egg from an ovary.
3. IVF stands for In Vitro Fertilisation, which is the process by which an egg is artificially fertilised by a sperm outside the body and an embryo re-implanted back into the uterus.
4. Oral contraceptives prevent the release of follicle stimulating hormone (FSH) and so stop ovulation happening, by preventing an egg developing.
5. To prevent the period (menstruation) happening and to maintain the uterus lining.
6. The high levels of oestrogen in the pill might produce side effects.

1.5 Responding to the environment

After studying this section you should be able to:

LEARNING SUMMARY

- explain the role of receptors and effectors in responding to stimuli
- describe the structure and function of the eye as an example of a sense organ
- recall the main eye defects.

Patterns of response

AQA	B1	✓
OCR A	B6	✓
OCR B	B1	✓
EDEXCEL	B1	✓
WJEC	B1	✓

All living organisms need to respond to changes in the environment. Although this happens in different ways the pattern of events is always the same.

KEY POINT

stimulus → detection → coordination → response

There are three main steps in this process:

- **Detecting the stimulus** – **receptors** are specialised cells that detect a stimulus. Their job is to convert the stimulus into electrical signals in nerve cells. Some receptors can detect several different stimuli, but they are usually specialised to detect one type of stimulus.

Stimulus	Type of receptor
Light	Photoreceptors in the eye
Sound	Vibration receptors in the ears
Touch, pressure, pain and temperature	Different receptors in the skin
Taste and smell	Chemical receptors in the tongue and nose
Position of the body	Receptors in the ears

KEY POINT

A sense organ is a group of receptors gathered together with some other structures.

The other structures help the receptors to work more efficiently. An example of this is the eye.

- Coordination – the body is receiving information from many different receptors at the same time.
  Coordination involves processing all the information from receptors so that the body can produce a response that will benefit the whole organism.

  In most animals this job is done by the central nervous system (CNS). The CNS is made up of the brain and spinal cord.

- Response – effectors are organs in the body that bring about a response to the stimulus. Usually these effectors are muscles and they respond by contracting. They could however be glands and they may respond by releasing an enzyme.

The eye – an example of a sense organ

OCR B	B1	✓
WJEC	B3	✓
CCEA	B1	✓

The structure of the eye.

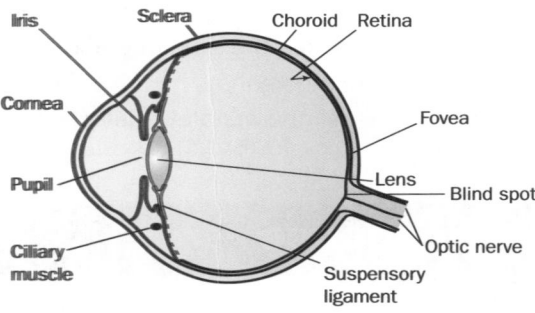

The light enters the eye through the pupil.

It is focused onto the retina by the cornea and the lens.

The size of the pupil can be changed by the muscles of the iris when the brightness of the light changes. The aim is to make sure that the same amount of light enters the eye.

The job of the lens is to change shape so that the image is always focused on the light sensitive retina.

The receptors are cells in the retina called rods and cones. They detect light and send messages to the brain along the optic nerve.

Remember that both the cornea and the lens bend, or refract, light, but it is the job of the lens to make the fine adjustment to focus the light on the retina.

Accommodation

| OCR B | B1 | ✓ |
| CCEA | B1 | ✓ |

The lens must be a different shape when the eye looks at a close object compared to a distant object. This is to make sure that the light is always focused on the back of the retina.

The ciliary muscle changes the shape of the lens as shown in the diagram.

This is called accommodation.

> To get an A*, you must be able to explain how accommodation occurs. Make sure that you remember that contracting the ciliary muscle allows the lens to become rounded and to refract light more.

How the eye focuses.

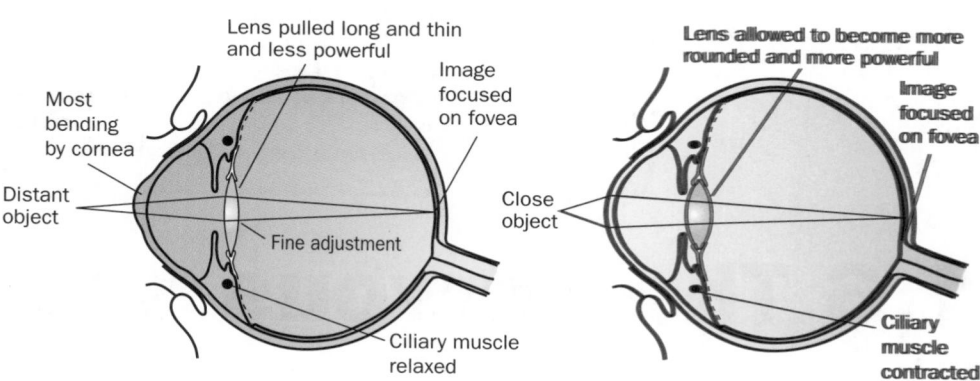

Some people have problems with their eyes. There are a number of different causes:

Condition	Cause	Treatment
Long-sight or short-sight	The eyeball or lens is the wrong shape.	Long-sight and short-sight can be corrected by wearing convex or concave lenses respectively. Cornea surgery can now also be used.
Red-green colour blindness	Lack of certain cones in the retina.	No treatment.
Poor accommodation	Lens becomes less elastic in senior citizens.	Wearing glasses with half convex and half concave lenses.

Judging distance

| OCR B | B1, B2 ✓ |

The eyes are also used to judge distances.

- Animals that hunt usually have their eyes on the front of their head. Each eye has a slightly different image of the object. This is called binocular vision and this can be used to judge distance.
- Animals that are hunted usually have eyes on the side of their heads. This gives monocular vision and they cannot judge distances so well. They can however see almost all around.

PROGRESS CHECK

1. What can receptors in the skin detect?
2. Muscles are the most common effector. Write down one other type of effector.
3. What is the job of the iris?
4. Why does a rabbit have eyes on the side of its head?
5. Explain why a convex lens is needed to correct long-sight.
6. Why does it cause eye strain to look at a close object for some time?

1. Receptors in the skin can detect touch, pressure, pain and temperature.
2. Glands are another type of effector.
3. The iris controls the size of the pupil/the amount of light entering the eye.
4. A rabbit has eyes on the side of its head so that it can see predators approaching from all angles.
5. Light rays are focusing behind the retina so they need to be bent/refracted more.
6. The ciliary muscles become tired due to being contracted for a long time.

1.6 The nervous system

LEARNING SUMMARY

After studying this section you should be able to:

- describe the structure and function of different neurones
- explain how a synapse works
- describe the role of neurones in reflexes and voluntary actions.

Nerves and neurones

AQA	B1	✓
OCR A	B6	✓
OCR B	B1	✓
EDEXCEL	B1	✓
WJEC	B3	✓
CCEA	B1	✓

To communicate between receptors and effectors the body uses two main methods. These are:

- nerves
- neurones.

KEY POINT

A neurone is a single, specialised cell that is adapted to pass electrical impulses.

The brain and spinal cord contains millions of neurones, but outside the CNS neurones are grouped together into bundles of hundreds or thousands. These bundles are called nerves.

The three main types of neurones are:

- Sensory neurones – they carry impulses from the receptors to the CNS.

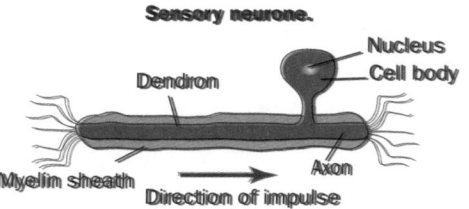

Sensory neurone.

Nucleus · Cell body · Dendron · Myelin sheath · Axon · Direction of impulse

- **Motor neurones** – they carry impulses from the CNS to the effectors.

Motor neurone.

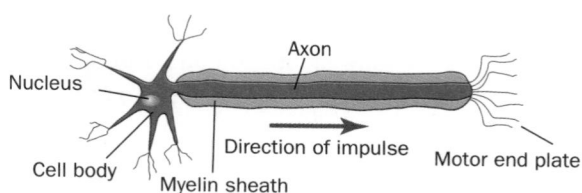

- **Relay neurones** – they pass messages between neurones in the CNS.

Although all neurones have different shapes, they all have certain features in common.

- One or more long projections (**axons** away from cell body and **dendrons** to the cell body) from the cell body to carry the impulse a long distance.
- A fatty covering (**myelin sheath**) around the projection which insulates it and speeds up the impulse.
- Many fine endings (**dendrites**) so that the impulse can be passed on to many cells.

Each neurone does not directly end on another neurone.

> **KEY POINT**
>
> There is a small gap between the two neurones and this is called a **synapse**.

Synapses

AQA	B1	✓
OCR A	B6	✓
OCR B	B1	✓
EDEXCEL	B1	✓
CCEA	B1	✓

So that an impulse can be generated in the next neurone, a chemical called a **neurotransmitter** is released when the nerve impulses reaches the synapse.

This then **diffuses** across the small gap and joins with receptors on the next neurone.

This starts a nerve impulse in this cell.

To get an A*, you must be able to explain why certain molecules might affect the working of synapses. Neurotransmitters have a specific shape to fit into receptors, so other molecules that have a similar shape might block the site or even act like the transmitter.

Chemical transmission between nerves.

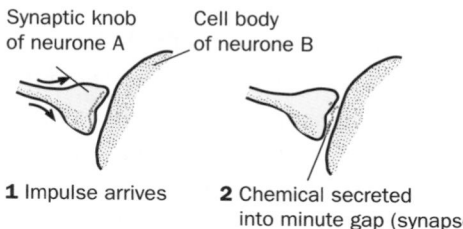

1 Impulse arrives   **2** Chemical secreted into minute gap (synapse)   **3** New impulse generated by neurone B

Many drugs work by interfering with synapses. They may block or copy the action of neurotransmitters in certain neurones.

Some of these are discussed on page 43.

Reflex responses

AQA	B1	✓
OCR A	B6	✓
OCR B	B1	✓
EDEXCEL	B1	✓
WJEC	B3	✓
CCEA	B1	✓

> **KEY POINT**
>
> The **peripheral nervous system** is made up of all the nerves that pass information to and from the CNS.

Once impulses reach the CNS from a sensory neurone there is a choice:

- Either the message may be passed straight to a motor neurone via a relay neurone. This is very quick and is called a **reflex action**.
- Or the message can be sent to the higher centres of the brain and the organism might decide to make a response. This is called a **voluntary action**.

All reflexes are:

- fast
- do not need conscious thought
- protect the body.

> Students often lose marks because they say that reflexes 'do not involve the brain'. Some do not, but some such as blinking do. They all do not involve conscious thought.

Examples of reflexes include the knee jerk, pupil reflex, accommodation, ducking and withdrawing the hand from a hot object.

This diagram shows the pathway for a reflex that involves the spinal cord:

A reflex action.

1 Stimulus is detected by sensory cell.

2 Impulse passes down sensory neurone.

3 Relay neurone passes impulse to motor neurone.

4 Motor neurone passes impulse to effector (muscle).

5 Muscle contracts.

> Use this short rhyme to remember the order of events in a reflex action:
>
> **Sue Remembers Seeing Rachel make Egg Rolls**
>
> **S**timulus, **R**eceptor, **S**ensory neurone, **R**elay neurone, **E**ffector, **R**esponse

PROGRESS CHECK

1. What is the difference between a nerve and a neurone?
2. What is the job of the fatty sheath around a neurone?
3. What is a synapse?
4. Why is it important to the body that reflexes are fast?
5. Scientists may want to produce a drug that is a painkiller. How can they use their knowledge of neurones and synapses to do this?
6. It is important that the body breaks down neurotransmitter molecules once they have stimulated a nerve impulse in the next neurone. Why is this?

1. A neurone is a single nerve cell, but a nerve contains thousands of neurones.
2. The fatty sheath around a neurone is there to insulate the neurone and speed up the rate of conduction.
3. A synapse is a small gap between two neurones.
4. It is important that reflexes are fast so that they can protect the body from damage.
5. Scientists could design a painkiller that could block the impulse travelling from the pain receptor, e.g. it might act on the synapses in the pathway, preventing the neurotransmitter from passing.
6. It is important that the body breaks down neurotransmitter molecules once they have stimulated a nerve impulse in the next neurone, otherwise they would stay in the receptor site and keep sending impulses. This could lead to convulsions.

1.7 Plant responses

LEARNING SUMMARY

After studying this section you should be able to:

- describe how plants respond to stimuli
- describe the roles of plant hormones in these responses
- describe how plant hormones can be used to manipulate plant growth.

Plant responses

AQA	B1	✓
OCR A	B5	✓
OCR B	B1	✓
EDEXCEL	B1	✓
WJEC	B1	✓
CCEA	B1	✓

Plants can also respond to changes in the external environment. These responses are usually slower than animal responses and include:

- roots and shoots growing towards or away from a particular stimulus
- plants flowering at a particular time
- the ripening of fruits.

> **KEY POINT**
>
> The type of response that involves part of the plant growing in a particular direction is called a **tropism**.

If the growth is in response to gravity it is a **geotropism (gravitropism)**.

If it is in response to light it is a **phototropism**.

Stimulus	Growth of shoots	Growth of roots
Gravity	Away = negatively geotropic.	Towards = positively geotropic.
Light	Towards = positively phototropic.	Away = negatively phototropic.

Remember it is important for shoots to grow away from gravity.

When a seed germinates it is often under the soil so the shoot cannot be growing towards light as it is dark!

By controlling the growth of plants, **auxins** (a type of **plant hormone**) can allow plants to respond to changes happening around them. This means that the roots and shoots of plants can respond to gravity or light in different ways.

These responses help the shoot to find light for photosynthesis and the root to grow down to anchor the plant in the soil and absorb water and mineral ions.

Plant hormones

AQA	B1	✓
OCR A	B5	✓
OCR B	B1	✓
EDEXCEL	B1	✓
WJEC	B1	✓
CCEA	B1	✓

Growth in plants is controlled by chemicals called plant growth substances or plant hormones. There are a number of different types, but the main types are called auxins.

Auxins are made in the tip of the shoot and move through the plant acting as the signal to make the shoot grow towards the light.

Auxins change the direction that roots and shoots grow by changing the rate that the cells elongate.

Auxins are responsible for making plant cells increase in length or elongate.

Experiments have shown that a series of steps are involved in the response:

- light is shone on one side of a shoot
- more auxin is sent down the side of the shoot that is in the shade
- this causes cells on the shaded side to elongate more
- the shoot therefore grows towards the light.

How auxins control tropisms.

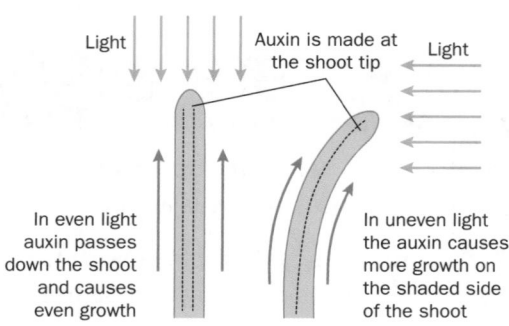

Light — Auxin is made at the shoot tip — Light

In even light auxin passes down the shoot and causes even growth

In uneven light the auxin causes more growth on the shaded side of the shoot

> To get an A*, you must be able to look at the results of experiments on shoots and work out which way they will grow. Learn and practise this skill by trying question 2 on page 31.

Applications of plant hormones

AQA	B1	✓
OCR B	B1	✓
EDEXCEL	B1	✓
CCEA	B1	✓

KEY POINT

Gardeners can use plant hormones such as auxins to help them control the growth of their plants.

Uses of auxins.

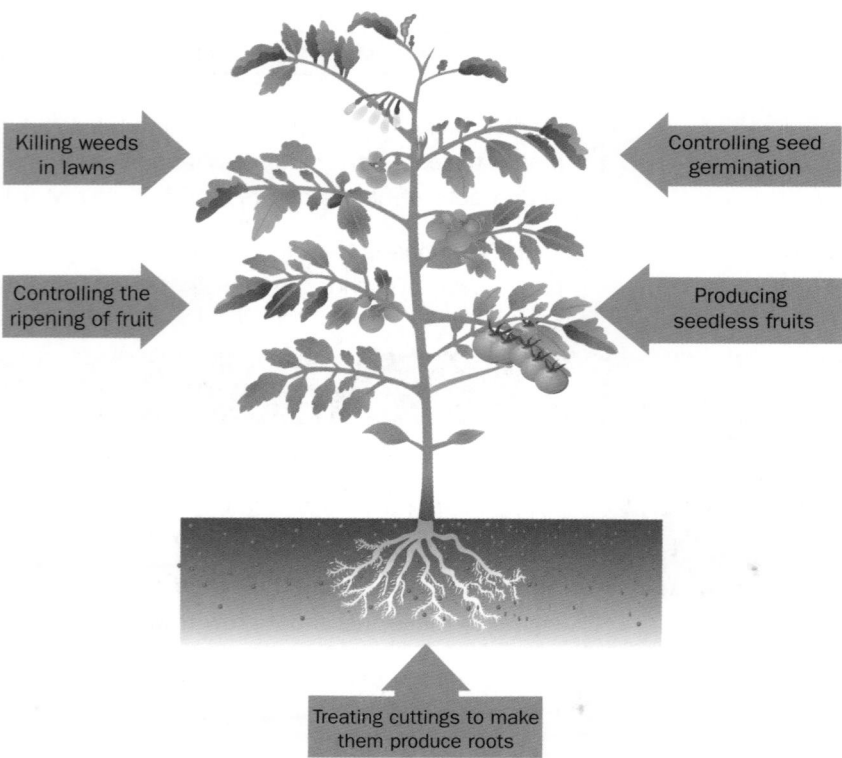

Killing weeds in lawns

Controlling seed germination

Controlling the ripening of fruit

Producing seedless fruits

Treating cuttings to make them produce roots

The plant hormones used by gardeners work in a number of ways:

- Auxins kill weeds because they make them grow too fast so they use up all their food reserves. These weed killers can be selective because they only kill broad leaf weeds, not grass.
- Some seeds need to be in the soil for some time before they germinate. Plant hormones can speed up this process.
- Fruit can be picked unripe so that it can be transported without being damaged. Then it can be ripened using plant hormones.

PROGRESS CHECK

1. Shoots are positively phototropic. What does this mean?
2. Roots are positively geotropic. Why is this important for a plant?
3. Where are auxins made in a plant?
4. Why are shoots dipped into hormone powder when taking cuttings?
5. Placing a foil cap over the tip of a shoot allows it to grow, but stops it bending towards light. Explain why.
6. Write down one advantage and one disadvantage of producing seedless fruits.

6. Advantage: more convenient to eat or more flesh to eat; Disadvantage: plant cannot reproduce sexually or cannot collect seeds to grow new plants.
5. The foil cap blocks the light, so auxin is evenly distributed in the shoot and so the shoot grows upwards.
4. Shoots are dipped into hormone powder to make them produce lateral roots.
3. Auxins are found in the shoot tip and root tip.
2. Positively geotropic means roots grow down into the soil anchoring the plant and finding water.
1. Phototropic means plants grow towards light.

Sample GCSE questions

1 The regulation of body temperature and water content of the body are examples of homeostasis.

(a) What is meant by homeostasis? **[2]**

This is the regulation of a constant internal environment in the body.

This is an important definition. Make sure that you learn it!

(b) Fill in this table to show ways that water is gained and lost by the body. **[5]**

Ways that water is gained	Ways that water is lost
drinking and eating	sweating
respiration	urine
	breathing

These are all correct points, although there are some others.

The one that is often missed is the production of water by respiration.

(c) The diagram shows sections through the human skin under different conditions.

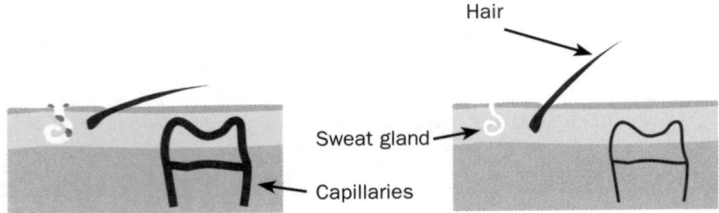

Hair

Sweat gland

Capillaries

Body temperature above normal   Body temperature below normal

A number of changes happen in the skin when the temperature falls below normal.

Describe and explain these changes.

The quality of written communication will be assessed in your answer to this question. **[6]**

The capillaries in the skin have closed down. This means that less blood flows close to the skin and so less heat is lost.

The sweat gland has also closed down so less sweat is lost. This would mean that less heat is lost by evaporation.

The hair stands on end. This is an attempt to trap more air close to the skin which would prevent heat loss.

Good explanation. Many candidates think that the capillaries move but they do not. You could use the term vasoconstriction to describe the closing.

The key word to use when talking about sweating is evaporation. This is what takes heat from the skin.

This still happens in humans but it only really has an effect in hairy mammals. You could mention that air is a good insulator.

Sample GCSE questions

(d) The graph shows the effect of changes in air temperature on sweat and urine production.

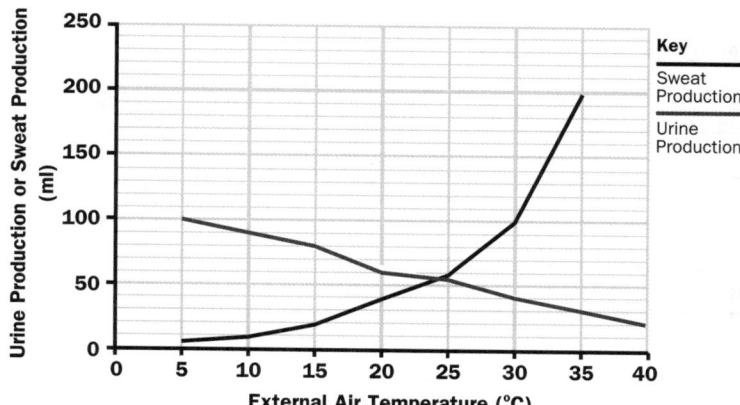

(i) Describe the patterns shown in the graph. **[2]**

As the external air temperature goes up the sweat production goes up and the urine production drops.

The urine production drops in a steady way but the rate of increase in sweat production increases.

(ii) Explain the changes shown by the graph. **[5]**

As it gets hotter the body temperature starts to increase. This is detected by the brain. The body responds with a number of changes aimed at losing more heat. One of these is to produce and release more sweat which evaporates from the skin.

The loss of water from the body makes the blood more concentrated and this is detected by the brain.

The body releases more of the hormone ADH and so less urine is produced and released.

This prevents the body getting too dehydrated.

This is a describe question so the answer needs to just relate to the graph.

This first mark is the easy mark describing the main difference.

Only the better candidates would give this point. The technical term for the shape of the sweat graph is exponential.

This is a good answer to a challenging question which links two topics together.

Remember only a fraction of a change in blood temperature will trigger changes so the blood temperature stays fairly constant.

You could give the site of ADH release which is the pituitary gland. It is also important to mention the kidneys which is where ADH acts.

Exam practice questions

1 The amount of each food substance needed by the body varies.

The table shows the EAR for four substances for different aged females.

These figures are per kilogram of body mass.

Age	Protein in g	Iron in mg	Vitamin C in mg
3 months	1.0	7	13
11 years	0.8	5	39
30 years	0.6	8	60
30 years and pregnant	0.9	22	70

(a) Which substance in the table is the most important for preventing scurvy?

.... **[1]**

(b) A 30 year-old woman has a mass of 60 kg.

She takes in 450 mg of iron in a day.

What effect might this have on her body?

Show your working.

....

.... **[2]**

(C) The EAR for protein for a 30 year-old woman is 0.6 times her body mass.

Compare this with the EAR figures for other females in the table, explaining any differences.

....

....

.... **[3]**

(d) The EAR for iron for an 11 year-old girl is about the same as that for an 11 year-old boy.

In the next few years this value becomes higher for the girl.

Suggest and explain a reason for this difference.

....

....

.... **[2]**

Exam practice questions

(e) The following information is on a packet of cereal.

Contents per serving	
Carbohydrate	68.7 g
Fat	14.0 g
Protein	9.0 g
Very low in salt content	trace amount

(i) A 30 year-old woman with a mass of 60 kg eats a serving of the cereal.

What percentage of her EAR will this supply?

....

.... **[2]**

(ii) The cereal is advertised as very low in salt.

Explain the health benefits of this.

....

....

.... **[2]**

2 Basil is doing an experiment to investigate how shoots respond to light.

He removes the tip from a growing shoot.

He places a jelly block containing a certain concentration of auxin on the cut end of the shoot as shown.

After a period of time Basil measures the angle of curvature of the shoot.

He repeats the process for different auxin concentrations.

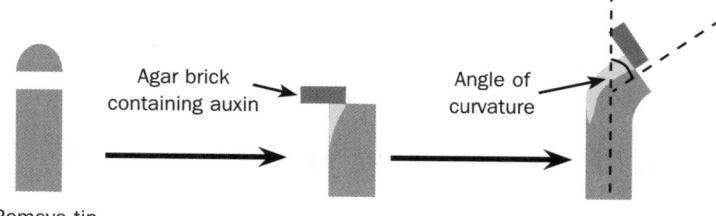

Remove tip — Agar brick containing auxin — Angle of curvature

Auxin concentration mg/dm³	0.05	0.1	0.15	0.2	0.25	0.30
Angle of curvature	4	9	13	19	22	17

Exam practice questions

(a) (i) Plot the results of the experiment on the graph.

Finish the graph by drawing the best curve. [3]

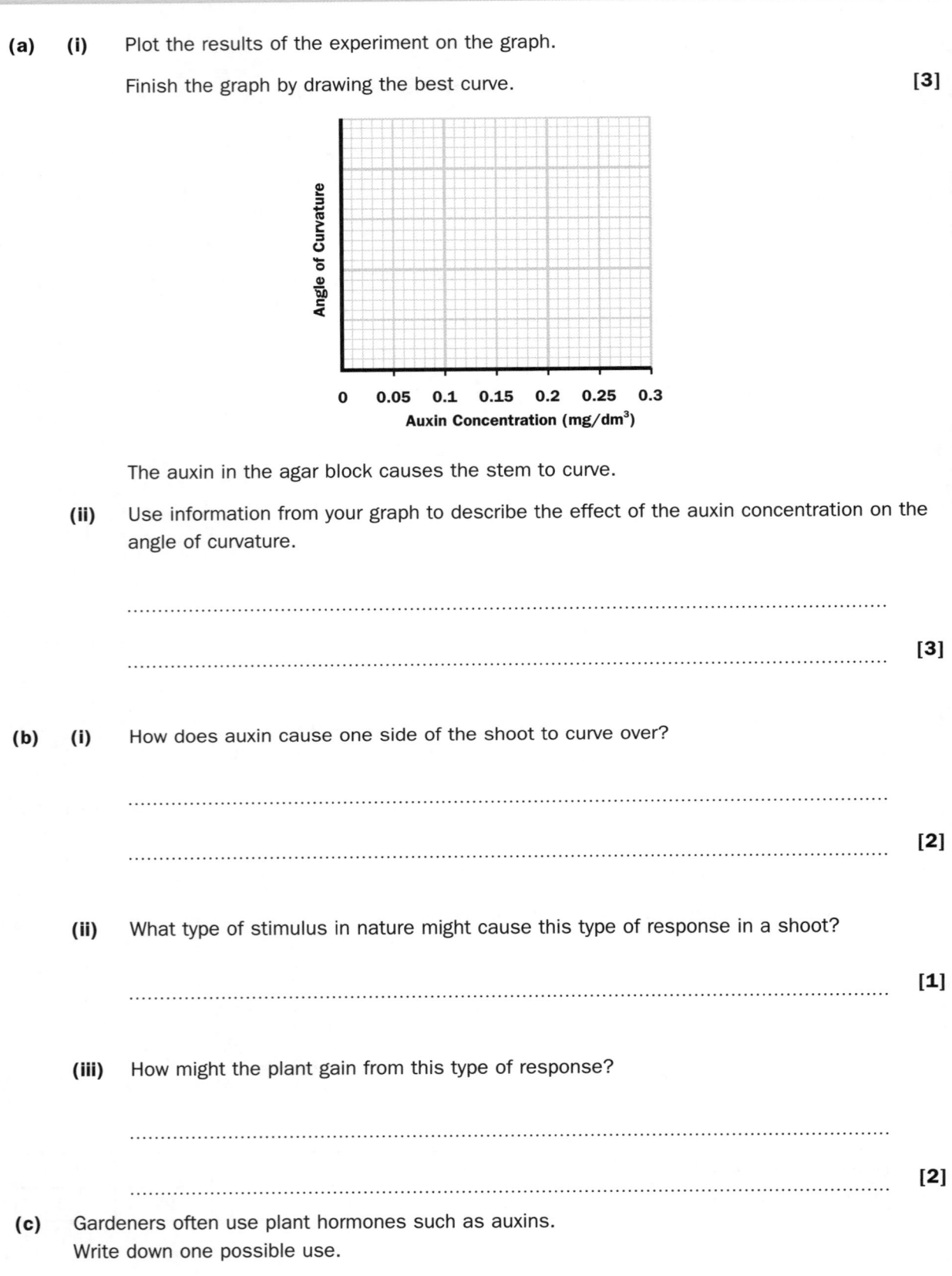

The auxin in the agar block causes the stem to curve.

(ii) Use information from your graph to describe the effect of the auxin concentration on the angle of curvature.

....

.... [3]

(b) (i) How does auxin cause one side of the shoot to curve over?

....

.... [2]

(ii) What type of stimulus in nature might cause this type of response in a shoot?

.... [1]

(iii) How might the plant gain from this type of response?

....

.... [2]

(c) Gardeners often use plant hormones such as auxins.
Write down one possible use.

.... [1]

Exam practice questions

3   The diagram shows a section through a human eye.

(a)   **(i)**   Label with **A** the part of the eye that carries nerve impulses to the brain.

   (ii)   Label with **B** the part of the eye that contains light sensitive cells.

   (iii)   Label with **C** the part of the eye that adjusts the size of the pupil.   **[3]**

(b)   The diagram shows light rays passing through the eye from a distant object.

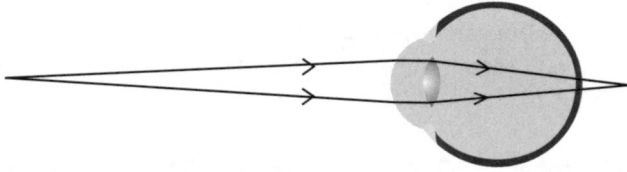

This person is long-sighted.

Explain why they cannot see the object clearly.

....

....   **[2]**

4   The diagram shows a motor nerve cell (neurone).

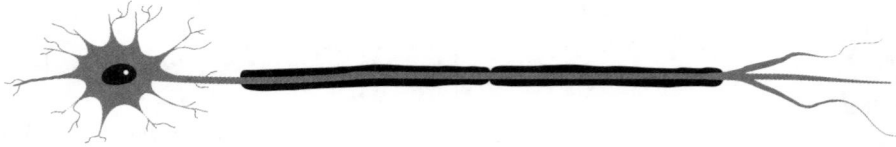

(a)   On the diagram label the fatty sheath.   **[1]**

(b)   Write the letter **A** on the part of the neurone that passes information onto a muscle.   **[1]**

(c)   Some people have a disease called motor neurone disease.
   Their motor neurones are gradually destroyed.
   These people can feel, smell and see perfectly normally but find breathing difficult.
   Explain the symptoms of this disease.

....

....

....   **[3]**

2 Health and disease

The following topics are covered in this chapter:

- Pathogens and infection
- Antibiotics and antiseptics
- Vaccinations
- Drugs
- Smoking and drinking
- Too much or too little

2.1 Pathogens and infection

LEARNING SUMMARY

After studying this section, you should be able to:

- recall some of the main causes of disease
- describe how pathogens can enter the body
- explain how white blood cells respond to antigens.

Causes of disease

AQA	B1	✓
OCR A	B2	✓
OCR B	B1	✓
EDEXCEL	B1	✓
WJEC	B3	✓
CCEA	B1, B2	✓

A disease occurs when the normal functioning of the body is disturbed. **Infectious diseases** can be passed on from one person to another, but **non-infectious diseases** cannot. Organisms that cause infectious diseases are called **pathogens**. There are a number of different types of organisms that can be pathogens.

> CCEA candidates need to know the main types of cancer, the main causes, methods for detection and treatment.

Type of disease	Description	Examples
Non-infectious		
Body disorder	Incorrect functioning of a particular organ	Diabetes, cancer
Deficiency disease	Lack of a mineral or vitamin	Anaemia, scurvy
Genetic disease	Caused by a defective gene	Red-green colour blindness
Infectious disease	Caused by a pathogen:	
	Fungi	Athlete's foot
	Viruses	Flu
	Bacteria	Cholera
	Protozoa	Malaria

Pathogens may reproduce rapidly in the body and may damage cells directly or produce chemicals called **toxins** which make people feel ill.

Viruses damage cells by taking over the cell and reproducing inside them. They then burst out of the cell destroying it in the process.

The entry of pathogens

OCR B	B6	✓
EDEXCEL	B1	✓
WJEC	B3	✓

There are a number of different ways that pathogens can be spread from one person to another.

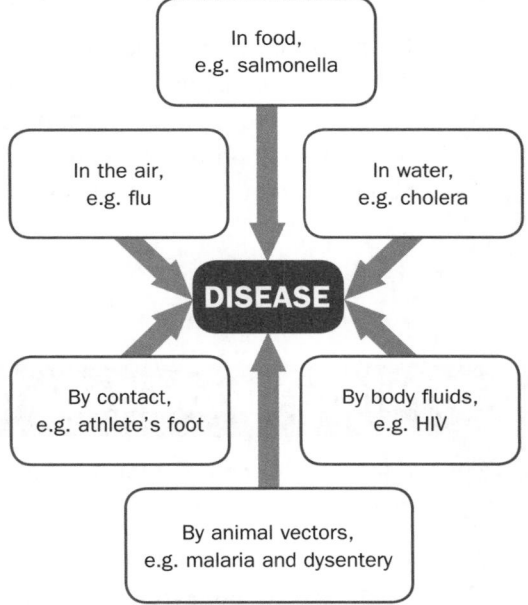

In food, e.g. salmonella

In the air, e.g. flu

In water, e.g. cholera

DISEASE

By contact, e.g. athlete's foot

By body fluids, e.g. HIV

By animal vectors, e.g. malaria and dysentery

The skin covers most of the body and is very good at stopping pathogens entering the body.

The body has a number of other defences that it uses in order to try to stop pathogens entering.

The body's defences.

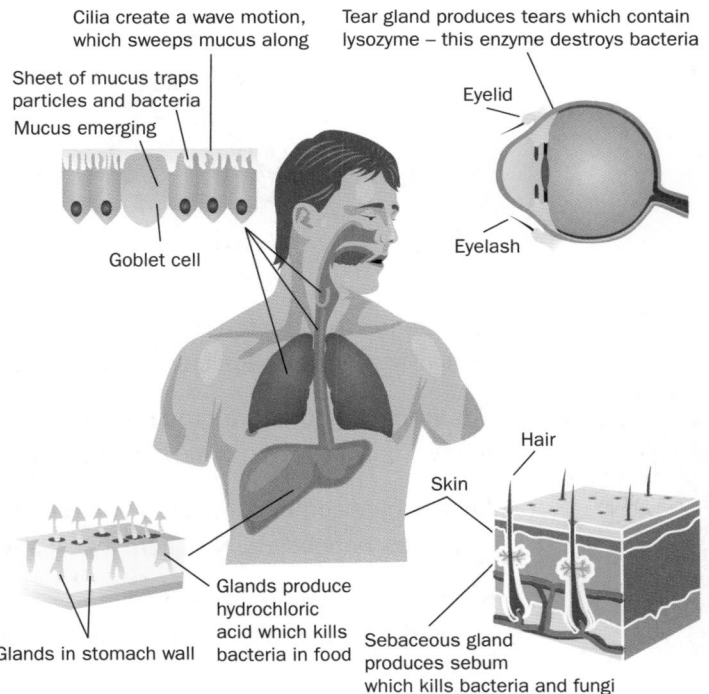

Cilia create a wave motion, which sweeps mucus along

Tear gland produces tears which contain lysozyme – this enzyme destroys bacteria

Sheet of mucus traps particles and bacteria

Mucus emerging

Eyelid

Goblet cell

Eyelash

Hair

Skin

Glands produce hydrochloric acid which kills bacteria in food

Glands in stomach wall

Sebaceous gland produces sebum which kills bacteria and fungi

Preventing the spread of disease

AQA	B1	✓
OCR A	B2	✓
OCR B	B1	✓
EDEXCEL	B1	✓
WJEC	B3	✓

By studying how pathogens are spread from person to person it is possible to find ways of preventing this spread. This will help to reduce the number of people getting the disease.

Make sure that you know which diseases are mentioned on your specification, which type of organism causes each disease and how they are spread from person to person.

For example, the malaria pathogen is spread by mosquitoes biting, therefore using insect repellents, insect nets and draining swamps where the mosquitoes breed can reduce malaria cases.

The action of white blood cells

AQA	B1	✓
OCR A	B2	✓
OCR B	B1	✓
EDEXCEL	B1	✓
WJEC	B3	✓
CCEA	B2	✓

If the pathogens do enter the body then the body will attack them in a number of ways.

The area that is infected will often become inflamed and two types of white blood cells (**phagocytes** and **lymphocytes**) attack the pathogen.

The actions of white blood cells.

Phagocyte   Lymphocyte

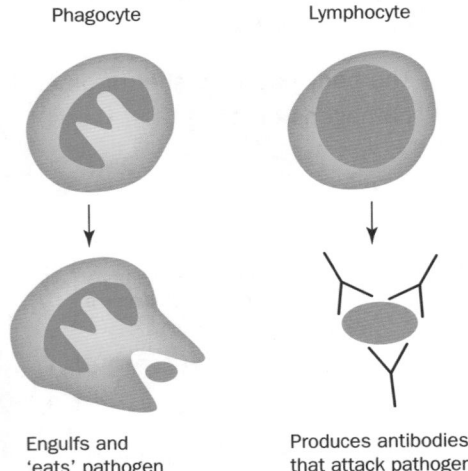

Engulfs and   Produces antibodies
'eats' pathogen   that attack pathogens

Pathogens are detected by the white blood cells because the pathogens have foreign chemical groups called **antigens** on their surface.

Students often lose marks because they confuse 'antibodies', 'antigens', 'antibiotics' and 'antiseptics'. Make sure you know the difference between them all! (Antibiotics and antiseptics are covered on the next page.)

The **antibodies** that are produced are specific to a particular pathogen or toxin and will only attach to that particular antigen.

When an antigen is detected by white blood cells they will produce memory cells as well as antibodies. The memory cells work by:

- living many years in the body
- producing antibodies very quickly if the same type of pathogen reinvades the body.

2.2 Antibiotics and antiseptics

LEARNING SUMMARY	**After studying this section, you should be able to:**
	• describe the difference between antibiotics and antiseptics
	• explain why some microbes are becoming resistant to antibiotics
	• describe how antibiotics and antiseptics are tested on microbes.

Antibiotics

AQA	B1	✓
OCR A	B2	✓
OCR B	B1	✓
EDEXCEL	B1	✓
WJEC	B3	✓
CCEA	B2	✓

Sometimes a pathogen can produce illness before our body's immune system can destroy it. It is sometimes possible for us to take drugs called **antibiotics** to kill the pathogen.

KEY POINT

Antibiotics:

- are chemicals that are usually produced by microorganisms, especially fungi
- kill bacteria and fungi but do not have any effect on viruses.

The first antibiotic to be widely used was penicillin, but there are now a number of different antibiotics that are used to treat different bacteria. This has meant that some diseases that once killed millions of people can now be treated.

There is a problem, however. More and more strains of bacteria are appearing that are resistant to antibiotics.

There are various ways that doctors try to prevent the spread of these resistant bacteria:

"I tell my patients to finish the dose of antibiotics even if they feel better."

"I change the antibiotics that I prescribe regularly and sometimes use combinations of different antibiotics."

"I prescribe antibiotics only in serious cases caused by bacteria."

"I always wash my hands with antiseptic between seeing patients."

Antibiotic resistance first appears due to a genetic change or mutation and soon afterwards a large population of resistant bacteria can appear.

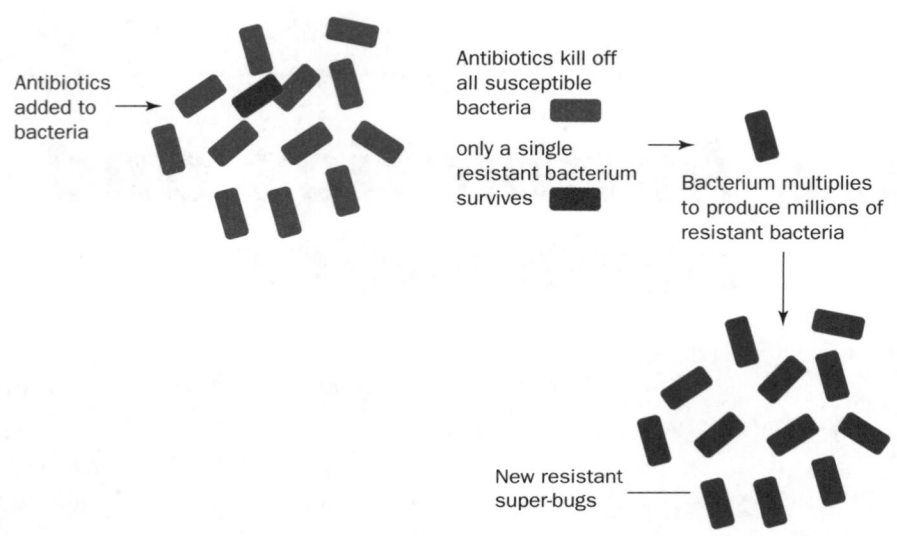

The development of antibiotic resistance.

Antibiotics added to bacteria

Antibiotics kill off all susceptible bacteria

only a single resistant bacterium survives

Bacterium multiplies to produce millions of resistant bacteria

New resistant super-bugs

This process has occurred in many different types of bacteria including the TB causing bacterium and one called MRSA. These bacteria are now resistant to many different types of antibiotic and so are very difficult to treat.

Antiseptics

AQA	B1	✓
OCR B	B6	✓
EDEXCEL	B1	✓

One important weapon against resistant bacteria is the use of **antiseptics**. Antiseptics:

- are man-made chemicals that kill pathogens outside the body
- were first used by an Austrian doctor called Dr Semmelweis to sterilise medical instruments
- are used widely in hospitals to try and prevent the spread of resistant bacteria.

An antiseptic is usually used on the body and a **disinfectant** is usually used on other surfaces.

> Draw a spider diagram in your revision book to show antibiotics, antiseptics, antibodies and disinfectants. Make sure it shows what they kill and what makes them.

Testing antibiotics and antiseptics

AQA	B1	✓
OCR A	B2	✓
OCR B	B1, B6	✓
EDEXCEL	B1	✓
WJEC	B3	✓
CCEA	B2	✓

It is possible to grow microorganisms such as bacteria in laboratories. They are grown on a special jelly called **agar** in a Petri dish. The agar is a culture medium containing an energy source, minerals and sometimes vitamins and protein.

Certain precautions have to be taken:

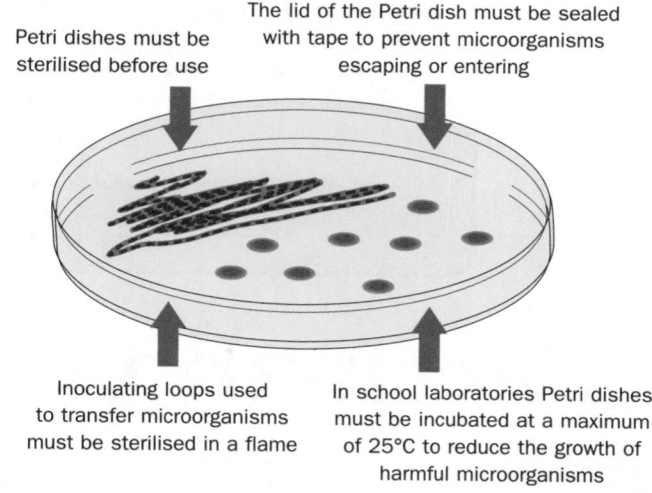

Petri dishes must be sterilised before use

The lid of the Petri dish must be sealed with tape to prevent microorganisms escaping or entering

Inoculating loops used to transfer microorganisms must be sterilised in a flame

In school laboratories Petri dishes must be incubated at a maximum of 25°C to reduce the growth of harmful microorganisms

It is then possible to see what action certain antibiotics or antiseptics have on the microorganisms.

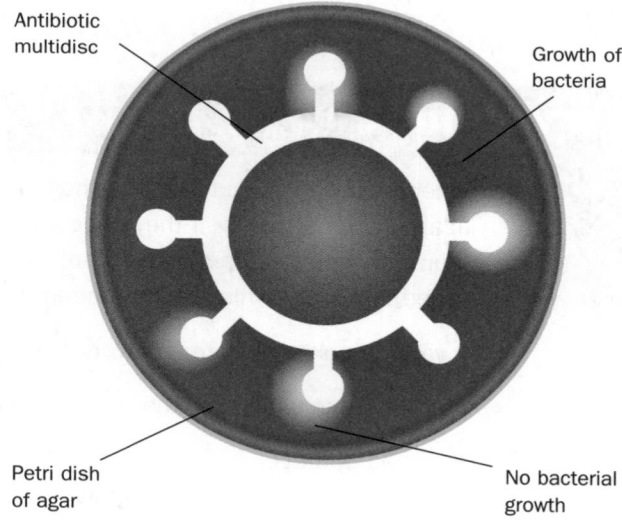

Antibiotic multidisc

Growth of bacteria

Petri dish of agar

No bacterial growth

Filter paper disks can be soaked in different antibiotics and placed on the agar.

The more effective the antibiotic, the wider the area of bacteria that will be killed.

2.3 Vaccinations

LEARNING SUMMARY

After studying this section, you should be able to:

- explain how a vaccine works
- explain the difference between active and passive immunity
- discuss some of the risks and advantages associated with vaccinations.

What is a vaccine?

AQA	B1	✓
OCR A	B2	✓
OCR B	B1	✓
EDEXCEL	B3	✓
CCEA	B2	✓

When our body encounters a pathogen, white blood cells make antibodies against the pathogen. If they encounter the same pathogen again in the future then antibodies are produced faster and the pathogen is killed quicker. This is called **immunity**. This idea has been used in **vaccinations**.

A vaccine contains harmless versions of the pathogen which stimulate immunity.

The harmless version of the pathogen contained in the vaccine could be:

- dead pathogens
- live, but weakened pathogens
- parts of the pathogen that contain antigens.

Questions often ask why it is necessary to produce a different flu vaccine every year. This is because different strains of the flu virus appear at regular intervals and they have different antigens. This means that the current memory cells would not recognise them.

Types of vaccine producing active immunity.

Bits of bacterial coat   Weakened virus

VACCINE

Dead bacteria

These all contain the specific antigens that are detected by the body's white blood cells. The memory cells that are produced stay in the body and will detect identical antigens in the future. This will lead to a more rapid immune response.

If a new strain of the pathogen appears then the current vaccination may not be effective.

Active and passive immunity

AQA	B1	✓
OCR A	B2	✓
OCR B	B1	✓
EDEXCEL	B3	✓
WJEC	B3	✓
CCEA	B2	✓

KEY POINT

The type of immunity, where the antibodies are made by the person, is called active immunity.

Sometimes it might be too late to give somebody this type of vaccination because they already have the pathogen.

KEY POINT

They can be given an injection containing antibodies made by another person or animal. This is called passive immunity.

It gives quicker protection but it does not last as long.

A vaccination containing antibodies.

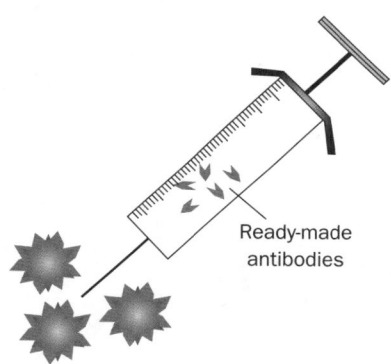

Ready-made antibodies

> Make a table to show the differences between active and passive immunity. Include who produces the antibodies, an example of when each might occur, how quickly they work and how long they last for.

Passive immunity also occurs when a baby receives antibodies from its mother across the placenta or in breast milk.

Vaccination risks and advantages

AQA	B1	✓
OCR A	B2	✓
OCR B	B1	✓
EDEXCEL	B3	✓

Whether or not to give your children vaccinations is a difficult decision to make for some people. Diseases like measles, mumps and rubella can have serious effects on the body.

- Measles is a very serious disease – 1 in 2500 babies that catch the disease die.
- Mumps may cause deafness in young children.
- Mumps may also cause viral meningitis which can be fatal.
- Rubella can cause a baby to have brain damage if its mother catches the disease during pregnancy.

The introduction of a combined **measles**, **mumps and rubella (MMR)** vaccine has led to a decrease in measles, mumps and rubella.

In 1998, a study of autistic children raised the question of a connection between the MMR vaccine and autism. (People with autism have difficulty with communicating and using some thinking skills.) This led to a decrease in the number of parents allowing their children to have the MMR vaccine. This study has been discredited, but it is impossible to say that having a vaccine does not involve a **risk**. Some people say that parents should be forced to allow their children to have the vaccine otherwise the disease will not disappear. Others say that it should be a personal choice.

PROGRESS CHECK

1. What does a vaccine contain?
2. What type of immunity is produced in the following cases:
   a) Antibodies are taken from a horse that has rabies and injected into a person.
   b) A person has chicken pox and is now immune to this disease.
3. Explain your answers to question 2.
4. Why do people sometimes feel ill after having certain vaccines?
5. Some pathogens can change their antigens during their life. What effect does this have?

5. People are not immune to the pathogen anymore, so the disease can reinfect.
4. If the vaccine contains live, but weakened, microbes then the person may get some symptoms of the disease.
3. In **a)** the antibodies have been made by a different organism so it is passive. In **b)** they have been made by the person so it is active.
2. **a)** Passive immunity.
   b) Active immunity.
1. Dead or weakened form of the pathogen.

2.4 Drugs

LEARNING SUMMARY

After studying this section, you should be able to:

- recall the main categories of drugs
- explain the effects that drugs can have on synapses
- discuss some of the issues arising from drug testing.

Types of drugs

AQA	B1	✓
OCR B	B1	✓
EDEXCEL	B1	✓
CCEA	B2	✓

Drugs are chemicals that alter the functioning of the body. Some drugs such as antibiotics are often beneficial to our body if used correctly. Others can be harmful, particularly those that are used recreationally.

Many drugs are **addictive**. This means that people want to carry on using them even though they may be having harmful effects. If they stop taking them they may suffer from unpleasant side effects called **withdrawal symptoms**. It also means that people develop **tolerance** to the drug, which means that they need to take bigger doses to have the same effect. Heroin and cocaine are very addictive.

Different drugs do different things.

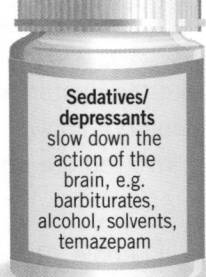  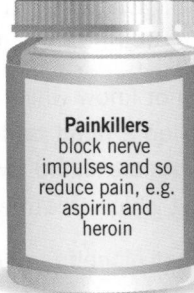 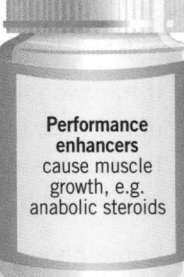 

Sedatives/ depressants slow down the action of the brain, e.g. barbiturates, alcohol, solvents, temazepam

Stimulants increase the activity of the brain, e.g. nicotine, ecstasy and caffeine

Painkillers block nerve impulses and so reduce pain, e.g. aspirin and heroin

Performance enhancers cause muscle growth, e.g. anabolic steroids

Hallucinogens distort what is seen or heard, e.g. cannabis and LSD

In order to control drugs many can only be bought with a prescription. Drugs are also classified into groups. Class A drugs are the most dangerous, with Class C being the least dangerous. If people are caught with illegal Class A drugs the penalties are the highest.

Drugs and synapses

| OCR B | B1 | ✓ |
| EDEXCEL | B1 | ✓ |

Stimulants such as nicotine affect synapses (the junction of two neurones) by causing more neurotransmitter substances to cross to the next neurone and bind to the receptor molecules. This makes it more likely for an impulse to be conducted in the next neurone.

> **KEY POINT**
>
> Depressants such as alcohol bind with receptor molecules, so blocking the transmission of impulses.

Testing new drug treatments

AQA	B1	✓
OCR A	B2	✓
OCR B	B1	✓
WJEC	B1	✓

Any new drugs have to be tested before they are used on patients. Doctors need to know:

- if the treatment works
- if it is safe.

There are a number of different ways that a new treatment can be tested:

- Firstly it is tested on cells in a laboratory. This is on human cells to see if it is harmful and on microorganisms in Petri dishes to see if it will kill them.
- If it passes these tests the drug is then tried on animals.
- Then the drug is tested on healthy human volunteers for safety and on people with the illness for safety and effectiveness.

> Exam questions often ask about testing drugs on animals or volunteers as this is an ideal 'How Science Works' subject. Be prepared to give both sides of the argument, even if you feel strongly one way or the other.

Many of these tests cause disagreements. Many people think that animals should not be used to test drugs. Some think that it is too cruel, while others think that it is pointless as the effects may be different on animals. Others think that the tests are reasonable because the benefits outweigh the risks of the tests.

Double–blind testing

AQA	B1	✓
OCR A	B2	✓
OCR B	B1	✓

Once a drug is cleared to be tested on patients the trial has to be set up carefully. One group is given the drug and another group has a **placebo**. A placebo looks like the real treatment, but has no drug in it.

> You must be able to analyse how a drugs test is done and decide if it is an open test, a blind test or a double–blind test. Look to see who knows which the real drug is.

If the two groups do not know which treatment they are having, but the doctor does, then this is called a **blind test**. If, in addition, the doctor does not know then this is called a **double–blind test**. It means that the people involved are not influenced by knowing which treatment is being given.

Some people think that placebos should not be used in tests on ill people. They say that it is not right to make people believe that they are receiving a possible cure when they are not.

Controversial drugs

| AQA | B1 | ✓ |

Over the years the use of some drugs has been particularly controversial.

Thalidomide is a drug that was given to pregnant women to try and relieve morning sickness. Thalidomide had been tested on pregnant animals. Unfortunately, many babies born to mothers who took the drug were born with severe limb abnormalities. The drug was then banned. More recently, thalidomide has been used successfully in the treatment of leprosy and other diseases.

Cannabis is an illegal drug. Many people have argued about whether it should be a Class B or Class C drug or possibly made legal. Cannabis smoke does contain harmful chemicals which may cause mental illness in some people.

PROGRESS CHECK

1. Why is it difficult to stop taking drugs such as cocaine?
2. Why are illegal drugs put into different classes?
3. Why are drugs tested on healthy volunteers?
4. Why do some people object to using animals to test drugs?
5. Before anaesthetics were available, surgeons often gave patients brandy to drink before operations. Suggest why they did this.
6. Explain why is it important that the doctor treating patients does not know whether the patients are taking the real drug or a placebo.

1. It is very addictive and changes the body chemistry, so that the body cannot function normally without it.
2. To indicate how dangerous they are and give guidance about punishments for illegal use.
3. To see if they have any side effects/test for safety.
4. They think that it is cruel/they may not have the same effect on animals.
5. Alcohol is a depressant. It will reduce synaptic transmission in neurones involved in pain.
6. So that the doctor can report the results without any bias and does not subconsciously interpret the same results in a different way.

2.5 Smoking and drinking

LEARNING SUMMARY

After studying this section, you should be able to:

- describe the effects of smoking and drinking alcohol on the body
- explain why smoking is particularly harmful during pregnancy.

Smoking

OCR B	B1	✓
EDEXCEL	B1	✓
WJEC	B2	✓
CCEA	B2	✓

KEY POINT

Many people cannot give up smoking tobacco because it contains the drug **nicotine**. This is addictive.

The nicotine is harmful to the body, but most damage is done by the other chemicals in the tobacco smoke.

Problems resulting from smoking.

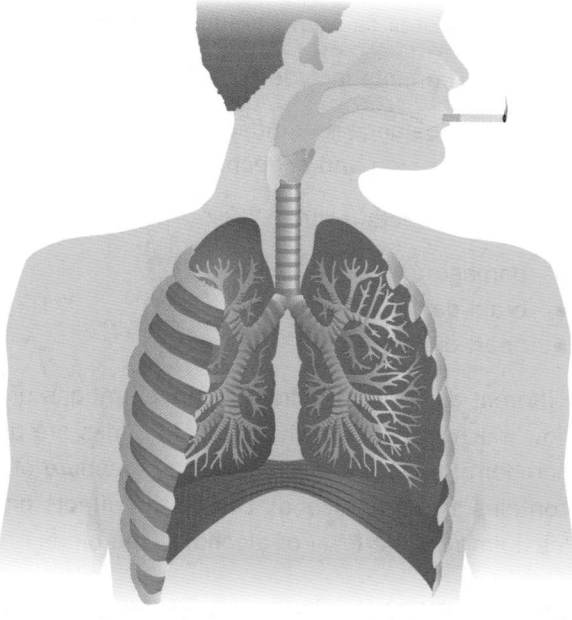

· The heat and chemicals in the smoke destroy the cilia on the cells lining the airways. The goblet cells also produce more mucus than normal. The bronchioles may become infected. This is called **bronchitis**.

· Chemicals in the tar may cause cells in the lungs to divide uncontrollably. This can cause **lung cancer.**

· The mucus collects in the alveoli and may become infected. This may lead to the walls of the alveoli being damaged. This reduces gaseous exchange and is called **emphysema**.

· The nicotine can cause an increase in blood pressure increasing the chance of a **heart attack**.

Smoking and blood pressure

OCR A	B2	✓
OCR B	B1	✓
CCEA	B2	✓

As well as the effects described in the diagram, smoking can increase blood pressure. It does this in two main ways:

- Nicotine increases the heart rate directly.
- **Carbon monoxide** reduces the oxygen-carrying capacity of the blood by combining with haemoglobin. This causes the heart rate to increase to compensate.

Smoking and pregnancy

| OCR B | B1 | ✓ |
| CCEA | B2 | ✓ |

Smoking tobacco is particularly dangerous for pregnant women. Mothers who smoke when they are pregnant are more likely to give birth to babies that have a low birth mass.

The graph shows a good example of a correlation. Although there is quite a spread in the data, the trend shows that the more cigarettes a mother smokes, the lighter her baby is likely to be. You may be given graphs showing similar trends with smoking and lung cancer or heart disease.

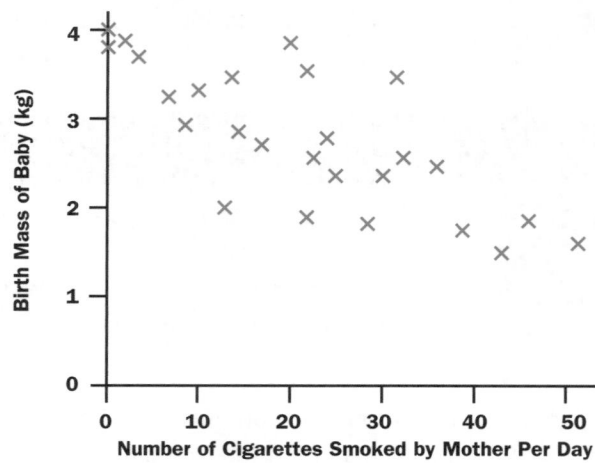

Drinking alcohol

OCR B	B1	✓
EDEXCEL	B1	✓
WJEC	B1	✓
CCEA	B2	✓

Drinking alcohol can have a number of effects on the body.

Short term effects include:

- loss of balance and muscle control
- blurred vision and speech.

Long term effects include:

- damage to the liver (cirrhosis)
- brain damage
- heart disease.

Different drinks have different concentrations of alcohol. To help people judge how much alcohol they have drunk, drinks are described as having a certain number of units of alcohol. A single measure of spirits, or half a pint of beer, contains 1 unit of alcohol. Due to the effects of alcohol on the body there is a legal limit for the level of alcohol in the blood of drivers and pilots.

Alcohol and reaction times

OCR B	B1	✓
EDEXCEL	B1	✓
CCEA	B2	✓

KEY POINT

Drinking alcohol increases reaction times.

This means that it is far more likely for drivers to have accidents if they have drunk alcohol recently. The graph on the following page shows this.

When answering questions about reaction times and stopping distances, you must remember that alcohol will increase reaction times and increase stopping distances. Candidates often get confused by this.

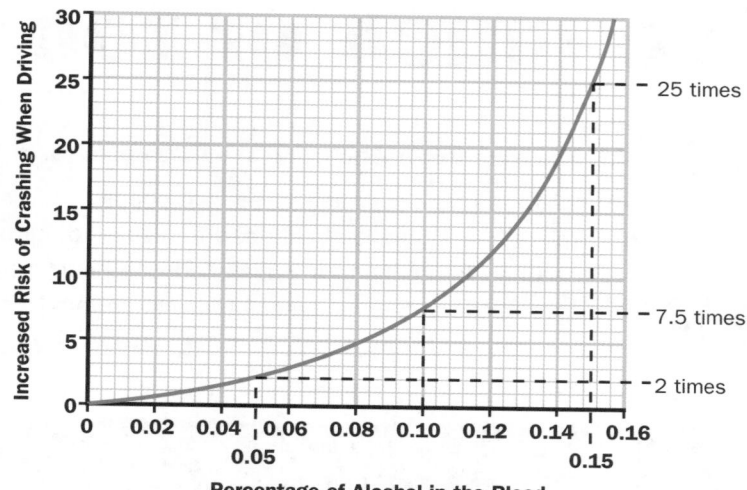

PROGRESS CHECK

1 What effect does smoking have on mucus production in the lungs?

2 Why is gaseous exchange reduced in emphysema?

3 What name is given to the damage caused to the liver by alcohol?

4 How many units of alcohol are there in two pints of beer and two single whiskeys?

5 If a person smokes, the oxygen content of the blood drops. Why is it necessary for their heart rate to increase to compensate?

6 How many times more likely is a person to have an accident if their blood alcohol level is 0.13% compared to 0.08%?

1. Increases production of mucus.
2. Walls of alveoli are broken down so reducing surface area for gas exchange. Alveoli are damaged so there is less gas exchange surface.
3. Cirrhosis.
4. Six.
5. So that blood is pumped faster around the body and so the cells receive sufficient oxygen to respire aerobically.
6. 15 times more compared with 5 times more – so an increase of 3 times.

2.6 Too much or too little

After studying this section, you should be able to:

- recall the problems associated with obesity
- explain how a poor diet can lead to high blood pressure
- describe the risk factors associated with heart disease
- describe the problems associated with certain restricted diets.

Eating too much

OCR A	B7	✓
OCR B	B1	✓
CCEA	B1	✓

It is important to maintain a balanced diet for the healthy functioning of the body. In the developed world many people eat too much food. This can make a person more likely to get various diseases. If a person eats food faster than it is used up by the body then the excess will be stored. Much of this will be stored as fat and can lead to **obesity**.

Obesity can be linked to a number of different health risks:

- arthritis – the joints wear out
- type 2 diabetes – unable to control the blood sugar level
- breast cancer
- high blood pressure
- heart disease.

It is possible to estimate if a person is underweight, normal, overweight or obese by using the formula:

> **KEY POINT**
>
> $$\text{Body Mass Index (BMI)} = \frac{\text{mass in kg}}{(\text{height in metres})^2}$$

The BMI figure can then be checked in a table to see what range a person is in.

Blood pressure

OCR A	B2	✓
OCR B	B1	✓
CCEA	B1, B2	✓

Contractions of the heart pump blood out into the arteries under pressure. This is so it can reach all parts of the body. Doctors often measure the blood pressure in the arteries and give two figures, for example 120 over 80. The highest figure is called the **systolic pressure** and this is the pressure when the heart contracts. The second figure is when the heart is relaxed and this is the **diastolic pressure**.

Blood pressure varies depending on various factors. The following factors can increase blood pressure:

- high salt and fat in the diet
- stress
- lack of exercise

- obesity
- high alcohol intake
- aging.

If left untreated, high blood pressure can cause various problems:

- Small blood vessels may burst, because of the high pressure. If a small blood vessel bursts in the brain, it is called a **stroke**. Brain damage from a stroke can result in some paralysis and loss of speech.
- If a small blood vessel bursts in a kidney, the kidney may be damaged.

Low blood pressure can cause problems such as:

- poor circulation
- dizziness and fainting, because the blood will not be at a high enough pressure to carry enough food and oxygen to the brain.

Heart disease

OCR A	B2	✓
OCR B	B1	✓
CCEA	B1, B2	✓

In heart disease the blood vessels supplying the heart muscle are blocked.

Lots of students lose marks because they say that fat blocks up blood vessels bringing blood back to the heart – do not make this mistake!

The heart is made up of muscle cells that need to contract throughout life. This needs a steady supply of energy so the cells need oxygen and glucose at all times for respiration. This is supplied by blood vessels.

Fatty deposits called **plaques** can form in these blood vessels and reduce the flow of oxygen and glucose to the heart muscle cells. This reduction in blood flow causes heart disease and if an area of muscle stops beating then this is a **heart attack**.

There are many factors that can make it more likely for a person to have heart disease.

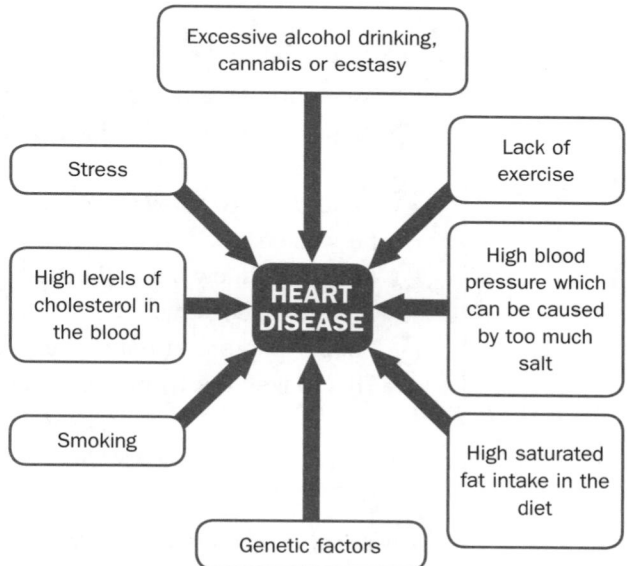

Most of these factors are lifestyle factors and managing these can reduce the risk of heart disease.

Coronary thrombosis

| OCR B | B1 | ✓ |
| CCEA | B2 | ✓ |

Plaques in the walls of the **coronary arteries** supplying the heart muscle make it more likely that blood will start to clot in the blood vessels. A blood clot inside the vessel is called a **thrombosis.** This blood clot may block the blood vessel. If the heart muscle does not get enough oxygen it will start to die. This leads to a heart attack.

Drugs such as **statins** can be taken to reduce the levels of cholesterol in the blood.

Eating too little

| OCR B | B1 | ✓ |

Eating too little of one type of food substance can lead to a deficiency disease.

Examples of deficiency diseases include:

- Anaemia due to a lack of iron.
- Scurvy due to a lack of vitamin C.
- Kwashiorkor due to a lack of protein.

Details of protein requirements and how to work out requirements are on page 10.

There are times when people do not eat enough food although there is food available. They may put themselves on a diet because they have a poor self image or think that they are overweight when they are not. This can reduce their resistance to infection and cause irregular periods in women. It may lead on to illnesses such as **anorexia**.

PROGRESS CHECK

1. A person has a mass of 80 kg and is 1.7 m tall. What is their BMI?
2. Why does a person's blood pressure have two figures?
3. What is a plaque?
4. Write down two ways that a person can try and reduce their risk of heart disease.
5. People with low blood pressure may often experience cold fingers and toes. Suggest why.
6. People who are at risk of heart disease take drugs such as warfarin. This makes the blood less likely to clot. Why do they take this drug?

6. Prevent a thrombosis happening in the coronary arteries which could block them and lead to a heart attack.
5. Poor circulation so limited blood supply to extremities which would usually bring heat.
4. Less stress/do not drink large amounts of alcohol/regular exercise/less saturated fat in diet/do not smoke.
3. A small build up of fat on the inside walls of the arteries.
2. The highest is the systolic when the heart contracts and the lowest is the diastolic when the heart relaxes.
1. $\frac{80}{1.7^2} = 27.7$

Sample GCSE questions

1 Read this newspaper article carefully and use the information to help you answer the questions.

TB bacteria may have met their match

Tuberculosis (TB) is a disease of the lungs.

TB is killing more people now for two main reasons.

Firstly, populations of the bacterium that cause TB are becoming resistant to many antibiotics.

Secondly, more people have the virus called HIV and this makes them much more likely to catch TB.

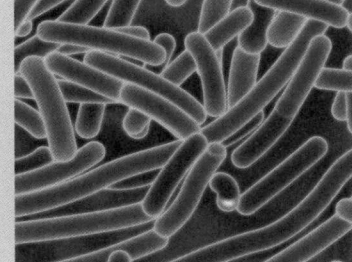

Scientists think that they have found a new antibiotic that could cure TB. The new antibiotic is being tested in double-blind tests.

(a) TB can usually be treated with antibiotics but HIV or the flu cannot be treated in this way. Explain why. **[2]**

HIV is a virus and flu is caused by a virus.

Antibiotics have no effect on viruses.

> This is correct but would probably only score one mark because it does not say that TB is caused by a bacterium and they are killed by antibiotics.

(b) Explain why antibiotic resistant populations of TB have appeared. **[3]**

Individual bacteria have been produced that have resistance to antibiotics due to mutations.

This means that when antibiotics were used all the non-resistant bacteria were killed leaving the resistant individuals to reproduce. They produced the resistant populations.

> A good answer. The two key ideas are mutations producing the resistance and these bacteria surviving to reproduce.

(c) The formation of resistant populations of bacteria can be slowed down.

Write down **one** way that doctors can help slow this process. **[1]**

Doctors can only prescribe antibiotics when they are really needed.

> This is a correct answer although 'when really needed' is a little vague. It is probably best to say 'only for serious bacterial infections'.

(d) The article says that the new drug will be tested in a double-blind test. What is involved in a **double-blind test**? **[3]**

This is when some patients get the drug and other patients get a blank drug.

Neither the patient nor the doctors giving the drugs know which treatment the patient is receiving.

> This answer is correct although the technical term for a blank drug is called a placebo and this could have been included.

Sample GCSE questions

(e) The first graph shows the percentage resistance to antibiotics of a type of bacteria called MRSA.

The second graph shows the number of new antibiotics that have been given approval to be used.

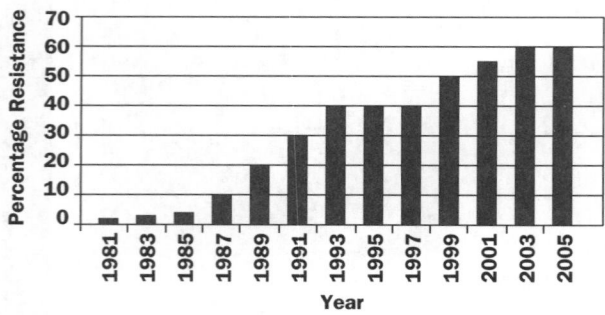

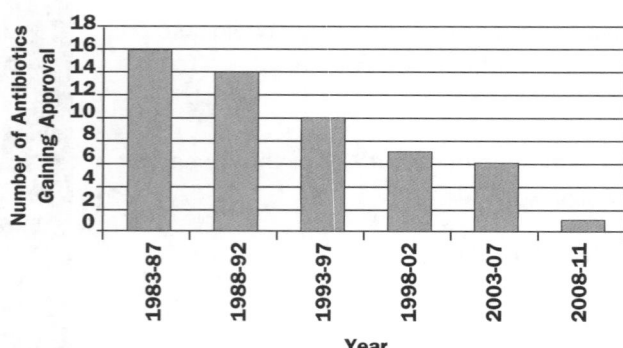

Use data from the two graphs to explain why scientists are concerned by antibiotic resistance in bacteria. **[3]**

Percentage resistance was less than 10% until about 1987, after which it increased dramatically. Although it was constant between 2003 and 2005 it was high at 60%.

The number of new antibiotics being approved is falling steadily. Only one new antibiotic was approved between 2008 and 2011.

To kill the resistant bacteria new antibiotics are needed.

They are not being produced.

This is a good answer, quoting figures from both graphs.
The trends in both graphs are described and there is a conclusion.

Exam practice questions

1 Vijay does not feel well so he visits the doctor.

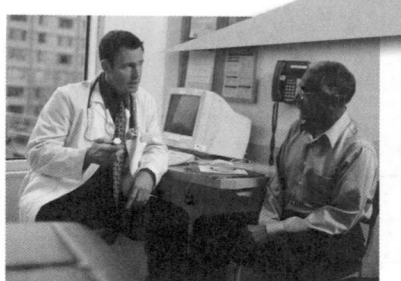

> Some bacteria have entered your body.
> They are producing chemicals that are making you feel ill.
> Your body will soon make cells that will kill the bacteria.
> To help kill the bacteria I will give you some medicine.
> If you had been given an injection when you were young
> then you would not have caught this disease.

(a) The boxes contain descriptions used by the doctor and certain biological terms.

Draw straight lines to join each **description** to the correct **biological** term.

description	biological term
The chemicals made by the bacteria that are making Vijay feel ill	Antibiotics
An injection that could have stopped Vijay getting the disease	Vaccination
The medicine given to Vijay to kill the bacteria	White blood cells
The cells produced by Vijay's body to kill the bacteria	Toxins

[3]

(b) The bacteria entered Vijay's body in his food.

Describe how bacteria in food are usually killed in the stomach.

.... [1]

(c) Write down **one** other method that the body uses to try and prevent bacteria entering the body.

.... [1]

(d) The doctor told Vijay that the medicine had been tested on animals.

Suggest why Vijay was pleased that the medicine had been tested.

.... [1]

Exam practice questions

2 The human lungs can be affected by various diseases.

One of these diseases is lung cancer.

In 1950 a scientist called Richard Doll investigated why people were getting lung cancer.

Richard Doll studied a large number of patients in London hospitals.

The table shows his observations.

Cause of death	Annual death rate per thousand patients		
	Non-smokers	Heavy smokers	All patients
lung cancer	0.0	1.1	0.7
all causes	13.6	16.3	14.0

(a) What conclusions can be made from Doll's data?

....

....

....

.... **[3]**

(b) Back in 1751, Dr John Hill observed several patients with cancer and wrote that tobacco may be the cause.

Richard Doll's ideas spread faster and were accepted by more people than John Hill's conclusions.

Suggest reasons for this.

....

....

.... **[2]**

3 **(a)** MMR is a vaccine that provides protection from three diseases in one injection.

About ten days after having the injection, children might get a measles-like rash.

Then, after three weeks they might get a mild form of mumps.

After six weeks, a rash of small spots like rubella may develop.

(i) People often feel ill after having a vaccination. Explain why.

....

.... **[2]**

Exam practice questions

(ii)   Some parents do not want their children to have the MMR vaccine.

They want their children to visit the doctor three times for three separate vaccinations.

Suggest why they may feel this way.

....

.... [2]

(b)   The Government has tried to persuade these parents to let their children have the MMR vaccination.

Suggest why the government want children to have one vaccination rather than three separate injections.

....

....

.... [2]

(c)   Explain how one injection can protect a child from three different diseases.

....

....

....

.... [3]

Exam practice questions

4 The graph shows the number of people dying from heart disease in Britain from 1928 to 1955.

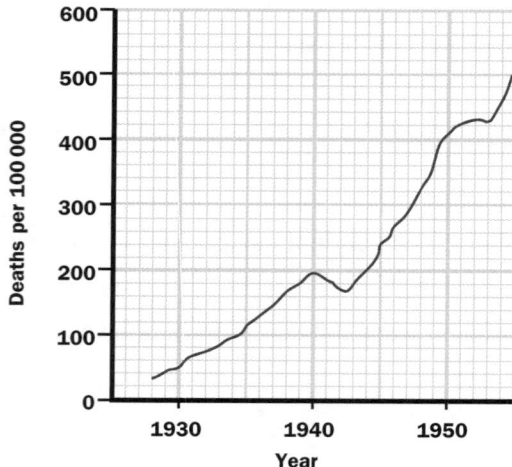

(a) Describe the pattern shown by the graph.

....

.... **[2]**

(b) The Second World War started in 1939.

During the war less food was available and people ate less fat.

Explain how the graph supports this fact.

....

....

.... **[3]**

(c) During the war farmers ate more fat than other people but doctors found that farmers had a lower rate of heart disease.

Suggest a reason why their rate of heart disease was lower than other people.

....

.... **[1]**

3 Genetics and evolution

The following topics are covered in this chapter:

- Genes and chromosomes
- Passing on genes
- Gene technology
- Evolution and natural selection

3.1 Genes and chromosomes

LEARNING SUMMARY

After studying this section, you should be able to:

- explain what is meant by the term gene
- describe the sources of variation produced by sexual reproduction
- explain how sex is determined
- discuss the importance of genes and the environment in variation.

What is a gene?

AQA	B1	✓
OCR A	B1	✓
OCR B	B1	✓
EDEXCEL	B2	✓
WJEC	B1	✓
CCEA	B2	✓

Most cells contain a nucleus that controls all of the chemical reactions that go on in the cell. Nuclei can do this because they contain the genetic material. Genetic material controls the characteristics of an organism and is passed on from one generation to the next.

> **KEY POINT**
>
> The genetic material is made up of structures called chromosomes. They are made up of a chemical called **deoxyribonucleic acid** or **DNA**.

The DNA controls the cell by coding for the making of proteins, such as enzymes. The enzymes will control all the chemical reactions taking place in the cell.

> **KEY POINT**
>
> A **gene** is part of a chromosome that codes for one particular protein.

DNA codes for the proteins it makes by the order of four chemicals called **bases**. They are given the letters **A, C, G** and **T**. By controlling cells, genes therefore control all the characteristics of an organism.

Different organisms have different numbers of genes and different numbers of chromosomes. In most organisms that reproduce by sexual reproduction, the chromosomes can be arranged in pairs. This is because one of each pair comes from each parent.

Chromosomes and reproduction

AQA	B1	✓
OCR A	B1	✓
OCR B	B1	✓
EDEXCEL	B2	✓
WJEC	B1	✓
CCEA	B2	✓

No living organism can live forever so there is a need to reproduce.

> **KEY POINT**
>
> **Sexual reproduction** involves the passing on of genes from two parents to the offspring.

> You only need to know the number of chromosomes in a human cell. Do not worry if a question asks about a different animal. Look for what information it supplies, for example it might say that a sperm of a fruit fly has four chromosomes. You can then work out that a leg cell would have eight.

This is why we often look a little like both of our parents. The genes are passed on in the **sex cells** or **gametes** which join at **fertilisation**. In humans, each body cell has 46 chromosomes in 23 pairs. This means that when the male sex cells (sperm) are made they need to have 23 chromosomes, one from each pair. The female gametes (eggs) also need 23 chromosomes. When they join at fertilisation it will produce a cell called a **zygote** that has 46 chromosomes again. This will grow into an embryo and a baby. This also means that the offspring that are produced from sexual reproduction are all different because they have different combinations of chromosomes from their mother and father.

Because the baby can receive any one of the 23 pairs from mum and any one of the 23 pairs from dad, the number of possible gene combinations is enormous. This new mixture of genetic information produces a great deal of variation in the offspring. This just mixes genes up in different combinations, but the only way that new genes can be made is by **mutation**. This is a random change in a gene.

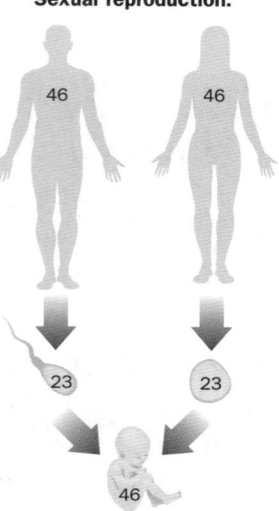

Sexual reproduction.

Sex determination

AQA	B2	✓
OCR A	B1	✓
OCR B	B1	✓
EDEXCEL	B3	✓
WJEC	B1	✓
CCEA	B2	✓

> **KEY POINT**
>
> In humans, the chromosomes of one of the 23 pairs are called the **sex chromosomes** because they carry the genes that determine the sex of the person.

There are two kinds of sex chromosome. One is called **X** and one is called **Y**.

* Females have two X chromosomes and are XX.
* Males have an X and a Y chromosome and are XY.

Females produce eggs that contain a single X chromosome and males produce sperm, half of which contain a Y chromosome and half of which contain an X chromosome. The diagram alongside shows the possible zygotes that can be produced by fertilisation.

The reason why the sex chromosomes determine the sex of a person is due to a single gene on the Y chromosome. This gene causes the production of testes rather than ovaries and so the male sex hormone testosterone is made. This will cause the development of all the male characteristics.

Variation

AQA	B1	✓
OCR A	B1, B3	✓
OCR B	B1	✓
EDEXCEL	B1	✓
WJEC	B1	✓
CCEA	B2	✓

Children born from the same parents all look slightly different. These differences are called **variation**. This can have different causes:

- **Inherited or genetic** – some variation is inherited from our parents in our genes.
- **Environmental** – some variation is a result of our environment.

Often our characteristics are a result of both our genes and our environment. The table shows examples of different kinds of variation.

Inherited	Environmental	Inherited and environmental
Earlobe shape	Scars	Intelligence
Eye colour	Spoken language	Body mass
Nose shape		Height
Dimples		

A good way to think of it, is that the genes provide a height and weight range into which we will fit, and how much we eat determines where in that range we will be.

Nature versus nurture

AQA	B1	✓
OCR A	B1	✓
OCR B	B1	✓

Scientists have argued for many years whether 'nature' or 'nurture' (inheritance or environment), is responsible for characteristics like intelligence, sporting ability and health. Some of the most important work on this subject has been done by studying identical twins that have been separated at birth.

> **PROGRESS CHECK**
>
> 1. What does DNA stand for?
> 2. What does DNA code for?
> 3. Why is it important that a sex cell has only one chromosome from each pair?
> 4. What mechanism can produce new genes?
> 5. Explain why approximately the same number of boys are born as girls.
> 6. Why are identical twins separated at birth so useful when studying nature versus nurture arguments?
>
> 1. Deoxyribonucleic acid.
> 2. Codes for proteins.
> 3. So that at fertilisation the full number of chromosomes can be restored.
> 4. Mutation.
> 5. Sperm are either X or Y in even numbers. If a Y sperm fertilises then it is a boy and an X sperm makes a girl.
> 6. Similarities are due to genetics (they have the same alleles) and so any differences must be due to the environment.

3.2 Passing on genes

LEARNING SUMMARY

After studying this section, you should be able to:

- explain the difference between the terms dominant and recessive
- explain the terms homozygous, heterozygous, genotype and phenotype
- construct genetic diagrams to predict the results of crosses
- recall the symptoms of certain genetic conditions
- discuss the ethical issues arising from genetic screening.

Different copies of genes

AQA	B2	✓
OCR A	B1	✓
OCR B	B1	✓
EDEXCEL	B1	✓
WJEC	B1	✓
CCEA	B2	✓

We have two copies of each chromosome in our cells (one from each parent). This therefore means that we have two copies of each gene. Sometimes the two copies are the same but sometimes they are different.

A good example of this is tongue rolling. This is controlled by a single gene and there are two possible copies of the gene, one that says roll and the other that says do not roll. If a person has one copy of each then they can still roll their tongue. This is because the copy for rolling is **dominant** and the non-rolling copy is **recessive**.

KEY POINT

Each copy of a gene is called an **allele**. If both alleles for a gene are the same this is called **homozygous**. **Heterozygous** means that the two alleles are different.

The only genes that cannot have two alleles present are those found on the X chromosome in men. This is because men only have one X chromosome. These genes are said to be **sex linked**.

The idea that characteristics were passed on as discrete 'factors' or genes was first suggested in 1866 by a monk called **Gregor Mendel**. At that time people believed that reproduction just caused factors to blend together. Using pea plants, Mendel showed that blending did not occur. At the time few scientists took any notice of his work because he was experimenting in a small monastery in the country. His work was rediscovered almost fifty years later.

We usually give the different copies (alleles) of a gene different letters, with the dominant copy a capital letter, for example T = tongue rolling and t = non-rolling.

Let us assume that Mum cannot roll her tongue, but Dad can. Both of Dad's alleles are T so he is homozygous. This is called his **genotype** as it describes what alleles he has. Rolling his tongue is called his **phenotype** as it describes the effect of the alleles. The cross is usually drawn out like this:

		Mum	
		t	t
Dad	T	Tt	Tt
	T	Tt	Tt

← All are tongue rollers

In this cross all the children can roll their tongue.

If both Mum and Dad are heterozygous the children that can produce will be different:

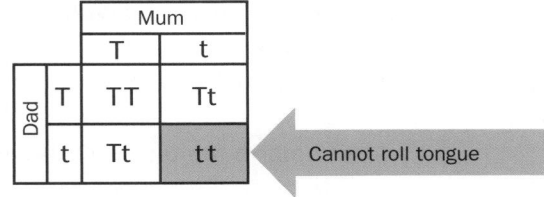

In this cross, 1 in 4 or 25% or a quarter of the children cannot roll their tongue.

> If you have to choose which letters to use in a cross, make sure that you use capital and small versions of the same letter and choose a letter that looks different in the two versions. Avoid C and c or S and s!

Genetic disorders

AQA	B2	✓
OCR A	B1	✓
OCR B	B1	✓
EDEXCEL	B1	✓
WJEC	B1	✓
CCEA	B2	✓

Many **genetic disorders** are caused by certain copies of genes. These can be passed on from mother or father to the baby and lead to the baby having the disorder. Examples of these disorders are cystic fibrosis, Huntington's disease and sickle-cell anaemia. People with these disorders become ill.

Cystic fibrosis	Huntington's disease	Sickle-cell anaemia
Caused by a recessive allele	Caused by a dominant allele	Caused by a recessive allele
Symptoms include: • thick mucus collects in the lung • breathing is difficult • chest infections • food is not properly digested.	Symptoms include: • muscle twitching (tremor) • loss of memory • difficulty in controlling movements • mood changes.	Symptoms include: • feeling tired or weak • coldness in the hands and feet • pain in the bones, lungs and joints

By looking at family trees of these genetic disorders and drawing genetic diagrams (such as the one for tongue rolling) it is possible for people to know the chance of them having a child with a genetic disorder. This may leave them with a difficult decision to make as to whether to have children or not.

Genetic screening

AQA	B2	✓
OCR A	B1	✓
OCR B	B1	✓
CCEA	B2	✓

Genetic cross diagrams can only work out the probability of a child being affected. It is now possible to test cells directly to see if they contain an allele for a particular genetic disorder. This is called **genetic screening**. This could be done at different stages:

- In an **adult**. This could tell the person if they are a carrier for the disorder and so if they may be able to pass it on. It could also tell if the person was going to develop a certain disorder later in life, for example Huntington's disease.
- In a **foetus**. Some cells can be taken from the foetus whilst the mother is pregnant. The parents can then find out if their baby will have the genetic disorder.
- In an **embryo** before it is implanted in the mother. If an embryo is produced by IVF outside the mother's body, then it can be tested before it is implanted in the mother. It is therefore possible to choose which embryos to put into the mother.

To get an A*, you must be able to describe arguments for and against genetic screening in a particular situation. Make sure you give both views.

The process of genetic screening brings with it some difficult ethical decisions:

- In an adult would you want to know if you were going to develop a disease from which there is no cure? Should your employer or your insurance company be told?
- In a foetus the parents could have to decide whether to have a termination or not.
- In an embryo the test is called **preimplantation genetic diagnosis**. Some people think that the destruction of early embryos is wrong. Others worry that the embryos may be tested and chosen for characteristics other than those involving disorders.

PROGRESS CHECK

1. Why can a person roll their tongue even if their cells have an allele for non-rolling?
2. Name a genetic disorder caused by a dominant allele.
3. What is the difference between genotype and phenotype?
4. What is genetic screening?
5. What is the difference between a gene and an allele?
6. Suggest why people may not want their insurance company or employer to have the results of their genetic screening.

1. The allele for rolling is dominant over the allele for non-rolling.
2. Huntington's disease.
3. Genotype is what alleles a person has and phenotype is how the alleles express themselves (the characteristics of the person).
4. Testing for a genetic disease.
5. A gene is a length of DNA that codes for a protein, an allele is a particular copy of a gene that codes for a particular variation of the protein.
6. They may not get insurance or the job they apply for if people know that they will develop a genetic disease in the future.

3.3 Gene technology

LEARNING SUMMARY

After studying this section, you should be able to:

- describe how plants can reproduce asexually
- describe how animals can be cloned
- describe the possible medical uses of stem cells
- discuss some of the uses and issues arising from genetic engineering

Cloning

AQA	B1	✓
OCR A	B1	✓
OCR B	B3	✓
EDEXCEL	B2	✓
WJEC	B1	✓
CCEA	B2	✓

KEY POINT

Bacteria, plants and some animals can reproduce **asexually**. This only needs one parent and does not involve sex cells joining.

All the offspring that are made are genetically identical to the parent.

Gardeners often use **asexual reproduction** to copy plants – they know what the offspring will look like.

Different organisms have different ways of reproducing asexually:

- The spider plant grows new plantlets on the end of long shoots.
- Daffodil plants produce lots of smaller bulbs that can grow into new plants.
- Strawberry plants grow long runners that touch the ground and grow a new plant.

Asexual reproduction produces organisms that have the same genes as the parent.

How Dolly was produced from a cloned cell.

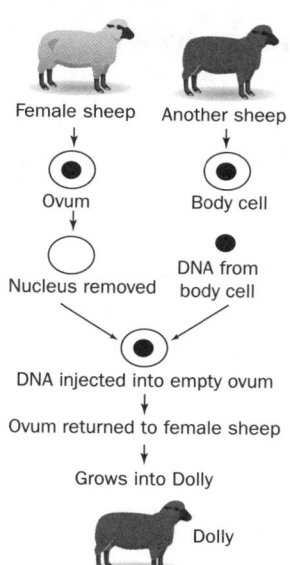

Female sheep → Ovum → Nucleus removed

Another sheep → Body cell → DNA from body cell

DNA injected into empty ovum

Ovum returned to female sheep

Grows into Dolly

Dolly

> **KEY POINT**
>
> Genetically identical individuals are called **clones**.

Many plants, such as the spider plant, clone themselves naturally and it is easy for a gardener to **take cuttings** to make identical plants. Modern methods involve **tissue culture** which uses small groups of cells taken from plants to grow new plants.

Cloning animals is much harder to do. Two main methods are used:

- **Cloning embryos** where embryos are split up at an early stage and the cells are put into host mothers to grow.
- **Cloning adult cells.** The first mammal to be cloned from adult cells was Dolly the sheep.

> Remember that clones have the same genes so any differences between them must be due to their environment.

Since Dolly was born other animals have been cloned and there has been much interest about cloning humans.

There could be two possible reasons for cloning humans:

- **Reproductive cloning** to make embryos for infertile couples.
- **Therapeutic cloning** to produce embryos that can be used to treat diseases.

Stem cells

AQA	B2	✓
OCR A	B1	✓
OCR B	B3	✓
EDEXCEL	B2	✓
WJEC	B2	✓

The use of embryos to treat disease is possible due to the discovery of **stem cells**.

> **KEY POINT**
>
> Stem cells are cells that can divide to make all the different tissues in the body.

They can be extracted from cloned embryos. Scientists think that they could be used to repair damaged tissues such as injuries to the spinal cord.

There are therefore many different views about cloning:

Both infertility and genetic disease cause much pain and distress. I think that we should be able to use cloning to treat these problems.

It is not right to clone people because clones are not true individuals and it is not right to destroy embryos to supply stem cells.

There are two main types of stem cells:

- **Embryonic stem cells** can develop into any type of cell. It is easy to extract them from an embryo, but the embryo is destroyed as a result.
- **Adult stem cells** can develop into a limited range of cell types. It is not necessary to destroy an embryo to get them, but they are difficult to find.

Genetic engineering

AQA	B1	✓
OCR B	B3, B6	✓
EDEXCEL	B2	✓
WJEC	B1	✓
CCEA	B2	✓

All living organisms use the same language of DNA. The four letters A, G, C and T are the same in all living things. Therefore a gene from one organism can be removed and placed in a totally different organism where it will continue to carry out its function. This means, for example, a cow will use a human gene to make the same protein that a human would make.

> **KEY POINT**
>
> Moving a gene from one organism to another is called **genetic engineering**.

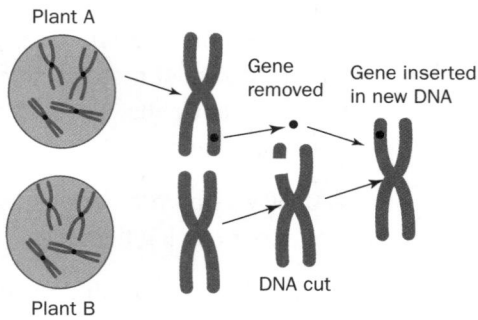

Genetic engineering.

There is also the possibility that genetic engineering may be used to treat genetic disorders like cystic fibrosis. Scientists are trying to replace the genes in people that have the disorder with working genes.

> **KEY POINT**
>
> Using genetic engineering to treat genetic disorders is called **gene therapy**.

GM crops

AQA	B1	✓
OCR B	B3	✓
EDEXCEL	B2	✓
WJEC	B1	✓

New **genetically modified (GM)** plants can be made in this way so that they:

- may be more resistant to insects eating them
- can be resistant to herbicides (weed killers)
- can produce a higher yield.

People often have different views about GM crops. Views against GM crops include:

- Genetic engineering is against 'God and Nature'.
- There may be long-term health problems with eating GM crops.
- Pollen from GM crops may spread to wild crops.

Views for GM crops include:

- More food to supply starving populations.
- Less need to spray harmful insecticides and herbicides.

PROGRESS CHECK

1. Why does a gardener use cuttings rather than seeds to reproduce an attractive plant?
2. What was Dolly?
3. What can stem cells do that normal body cells cannot?
4. Write down one characteristic that is chosen for GM crops.
5. Many people think that using adult stem cells to treat disease is acceptable, but are against using embryonic stem cells. Suggest why this is.
6. Suggest why farmers might want a crop that is resistant to herbicides.

1. Cuttings will produce an identical copy so the gardener can be sure of the characteristic of the plant.
2. Dolly was a sheep and the first mammal that was produced by cloning from an adult cell.
3. They can differentiate into any other type of cell.
4. Resistance to insects eating them/resistance to herbicides/produce a higher yield.
5. Use of embryonic stem cells involves the destruction of an embryo, but using adult stem cells does not. Some people consider an embryo to be an individual life.
6. They can spray their whole field with weedkiller, killing all the weeds, except the crop.

3.4 Evolution and natural selection

LEARNING SUMMARY

After studying this section, you should be able to:

- recall the meaning of the term evolution
- explain Darwin's theory of natural selection
- apply natural selection to recent examples of population changes
- compare Darwin's theories to those of Lamarck.

Evolution and fossils

AQA	B1, B2	✓
OCR A	B3	✓
OCR B	B2	✓
EDEXCEL	B1	✓
WJEC	B1	✓
CCEA	B2	✓

Most scientists now think that life on Earth started about 3500 million years ago.

How life started and why there is such a great variety of organisms are questions that people have argued over for a long time.

In the 1800s scientists questioned more about what **fossils** were.

Fossils are the remains of organisms from many years ago. Many early life forms did not fossilise because they were soft bodied. However, fossils can be formed in a number of ways:

- From the hard parts of organisms that do not decay.
- From parts of organisms that do not decay because conditions for decay are absent.

An example of the use of fossils is the tracing of the development of the five digits present in all vertebrates. This is called the pentadactyl limb.

- When parts of organisms are replaced by other substances as they decay.
- As preserved traces of organisms, e.g. footprints.

Many people at that time believed in creation. They said that organisms were created as they exist now, by God.

However, scientists found fossils of organisms such as dinosaurs that are not alive today. Some people started to believe the idea that species of organisms could gradually change.

> **KEY POINT**
>
> **Evolution** is the gradual change in a species over a long period of time.

The problem for the believers in evolution was that at first they could not explain how the gradual changes happened.

Charles Darwin

AQA	B1	✓
OCR A	B3	✓
OCR B	B2	✓
EDEXCEL	B1	✓
WJEC	B1	✓
CCEA	B2	✓

Charles Darwin (1809–1882) was a naturalist on board a ship called the HMS Beagle. His job was to make a record of the wildlife seen at the places the ship visited.

On his travels, Darwin noticed four things:

- Organisms often produce large numbers of offspring.
- Population numbers usually remain constant over long time periods.
- Organisms are all slightly different – they show variation.
- This variation can be inherited from their parents.

Darwin used these four simple observations to come up with a theory for how evolution could have happened. Darwin said that:

- All organisms are slightly different.
- Some are better suited to the environment than others.
- These organisms are more likely to survive and reproduce.
- They will pass on these characteristics and over long periods of time the species will change.

> You need to be able to use Darwin's theory to explain how a group of organisms has evolved. You may not have heard of the organisms before, but just use these main points:
>
> variation → best adapted survive → reproduce → pass on genes.

> **KEY POINT**
>
> Darwin called this theory **natural selection**.

> When explaining how natural selection happens, remember to talk about groups or populations of organisms changing over time. One organism does not evolve; it either survives to reproduce or it dies.

Darwin was rather worried about publishing his ideas. When he finally published them they caused much controversy. Many people were very religious and believed in creation. It took many years before Darwin's theory was generally accepted.

Because natural selection takes a long time to produce changes it is very difficult to see it happening. One of the first examples to be seen was the peppered moth. This moth is usually light coloured, but after the Industrial Revolution a black type became common in polluted areas.

This can be explained by natural selection:

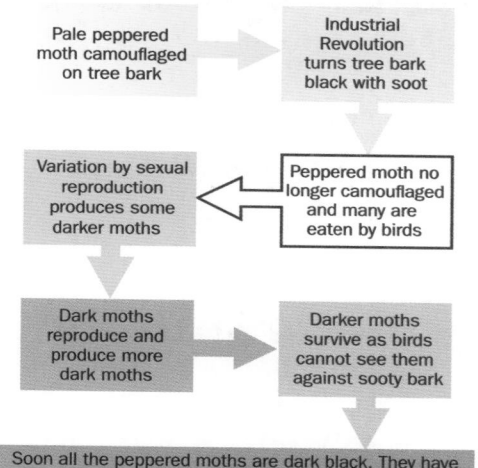

Other examples that can be explained by natural selection include:

- Rats becoming resistant to the rat poison warfarin.
- Bacteria becoming resistant to antibiotics.

Different theories for evolution

AQA	B1	✓
OCR A	B3	✓
OCR B	B2	✓

To get an A*, you must be able to spot an explanation of evolution made using Lamarck's ideas and explain why it is now thought to be wrong. An example might be that giraffes have long necks because they stretched to reach higher leaves and every generation they grew a little longer.

Darwin was not the first person to try and explain how evolution may have happened. A French scientist called **Lamarck** (1744–1829) said that organisms were changed by their environment during their life. They then passed on the new characteristics and so the population would change.

Darwin's and Lamarck's ideas are **theories** that explain **data** known at the time. As different data becomes known then people often start to accept different theories.

Most scientists now think that Lamarck's theory is wrong, because we now know that characteristics are passed on in our genes and genes are not usually altered by the environment.

Most people now accept Darwin's theory because it best explains all the data that has been discovered. However, it is only a theory, it is not fact.

PROGRESS CHECK

1. What does the word extinct mean?
2. Write down one organism that has been investigated by looking at fossils.
3. What advantage did the dark moths have in industrial areas?
4. Why was Darwin worried about publishing his ideas?
5. The terms 'struggle for survival' and 'survival of the fittest' are often used when describing Darwin's ideas. Explain what they mean.

5. The best adapted organisms are the ones that are most likely to survive in an environment.
4. Darwin thought that religious people would be against his ideas because his theory went against the beliefs at the time.
3. They were camouflaged against the polluted buildings and trees, so less likely to be predated on.
2. Any dinosaur species.
1. All the individuals of a species no longer exist.

Sample GCSE questions

1 There are about five million different types of organisms living on Earth.

A number of different theories have been put forward to try and explain how this variety has come about.

Here are three different theories:

Creation theory says that the Earth and all life on it were created by God as described in the Bible. Only small changes have happened since creation and no new species have been created.

Darwin said that all organisms were slightly different. Those organisms that were better suited would pass on their characteristics and so the population would gradually change.

A French scientist called **Lamarck** said that organisms were changed by their environment during their life. They then passed on the new characteristics and so the population would change.

(a) Which one of these three theories does **not** include evolution in its explanation?

Explain your answer. **[2]**

This is creation. Both of the other theories say that organisms have changed considerably over time. In creation they have not.

Remember that Darwin was not the only person that put forward ideas about evolution.

(b) Why is Darwin's theory still called a theory even though most scientists believe it to be true? **[1]**

It has not been proved correct.

This is an important point. These are all theories that explain data. Darwin's theory is now accepted by many because it best explains the data. It is not fact.

(c) **(i)** People often explain the difference between Lamarck's theory and Darwin's theory by using the example of the long necks of giraffes.

Lamarck's theory would explain this by saying that the necks of giraffes have stretched slightly during life to reach higher leaves.

This increase in length is passed on and after many generations the necks are longer.

Write down the explanation that would be given by Darwin's theory. **[4]**

Giraffes all show variation and so have different length necks. Those with the longer necks can reach more leaves and so get more food. They are more likely to survive and pass on this characteristic. After many generations, the necks of the giraffes are longer.

A good answer but you could include the name of the theory i.e. natural selection. A more modern version of natural selection would talk about mutations and genes but Darwin did not know about them.

Sample GCSE questions

(ii) Explain why most scientists do not believe Lamarck's theory. **[2]**

Lamarck's theory says that characteristics gained during their lifetime are passed on. Characteristics are passed on by genes and these changes would not alter the genes.

(d) After Darwin published his theory this cartoon appeared in an important magazine.

> The characteristics in Lamarck's theory are often called 'acquired characteristics'.

PUNCH, OR THE LONDON CHARIVARI.—MAY 25, 1861.

THE LION OF THE SEASON.

ALARMED FLUNKEY. "MR. G-G-G-O-O-O-RILLA!"

Write about why people were upset enough to publish this cartoon and what they were trying to show. **[4]**

Many people at that time were very religious and believed in creation. They did not want to believe in evolution. They were making fun of Darwin's ideas and suggesting that Darwin said that humans are descended from apes.

> Remember Darwin never said that humans were descended from apes or monkeys. Humans and apes share a common ancestor.

Exam practice questions

1 The photo shows a type of plant called a spider plant.

One plant can reproduce on its own by growing new plants at the end of shoots.

(a) What type of reproduction is this plant showing?

.... **[1]**

(b) Write down two advantages to a gardener of producing plants by this type of reproduction rather than by using seeds.

1 ....

....

2 ....

.... **[2]**

(c) New plants can also be produced by tissue culture.

Describe **one** difference between tissue culture and the type of reproduction shown by the spider plant.

....

.... **[1]**

2 Read the following passage carefully and use it to help you answer the questions.

> Albinism is a condition that causes a person to produce very little coloured pigment in their skin, hair or iris. This means that they have white hair, pink irises and very pale skin.
>
> The condition is not usually life threatening if the person takes some simple precautions.
>
> Albinism causes the person to become sun burnt very easily and the action of ultraviolet light on the skin is more likely to cause mutations.
>
> The condition is caused by a recessive allele (a). The dominant allele (A) causes normal pigment production.

(a) The action of ultraviolet light on the skin causes mutations.

What are mutations?

.... **[1]**

Exam practice questions

(b) Suggest one 'simple precaution' that a person with albinism might take.

.... **[1]**

(c) The allele that causes albinism is said to be recessive.

What does this mean?

....

....

.... **[2]**

(d) Explain how two parents who **do not** have albinism can produce a child that **does** have albinism.

Use a genetic diagram to help you.

....

....

.... **[4]**

3 **(a)** Peter and Kirsten are expecting a baby.

They know that it has an even chance of being a boy or a girl.

Finish the genetic diagram to show why this is.

Kirsten

Gametes	X	X

Peter

[2]

Exam practice questions

(b) The table shows the ratio of males to females in different countries and at different ages.

Age	India	Kenya	Russia	UK
at birth	1.12	1.02	1.06	1.05
over 65 years old	0.91	0.83	0.44	0.76
all ages	1.08	1.01	0.85	0.98

(i) In which countries do women live longer than men?

Use data from the table to justify your answer.

....

....

.... **[2]**

(ii) In some countries parents want to have baby boys rather than girls.

Embryos can be tested to see what sex they are before they are born.

People are worried that this technique might be used to terminate female embryos.

In which country would the data in the table suggest this is possibly happening?

Use data from the table to justify your answer.

....

....

.... **[2]**

(iii) Suggest **one** reason why people might be concerned about the effect on the country of the termination of female embryos.

....

.... **[1]**

Exam practice questions

4 Scientists believe that man evolved from ape-like animals several million years ago.

One important change was in the bones which allowed our ancestors to walk upright on two feet.

Scientists think that several millions of years ago the Earth became drier and forests were replaced by grasslands. Before this time all apes walked on four feet. In the grassland, populations of apes developed to walk on two feet. This enabled the ape to see further.

(a) Suggest why it might be an advantage for the ape to be able to see further.

.... **[1]**

(b) Explain how the ape developed so that it walked upright.

Use ideas about natural selection in your answer.

The quality of written communication will be assessed in your answer to this question.

....

....

....

.... **[5]**

(c) Another theory for why humans became upright involves the use of tools.

It suggests that standing on two feet allows the other two limbs to handle tools.

This needs a large brain to control the hands.

Recently a fossil has been found of a human ancestor that has the bones of an upright animal but a small brain.

Explain what this find indicates.

....

....

.... **[2]**

4 Organisms and environment

The following topics are covered in this chapter:

- Classifying organisms
- Competition and adaptation
- Living together
- Energy flow
- Recycling
- Pollution and overexploitation
- Conservation and sustainability

4.1 Classifying organisms

LEARNING SUMMARY

After studying this section, you should be able to:

- describe the principles of the modern classification system
- explain how species are defined
- describe the classification of the vertebrates
- describe how organisms are named.

Classifying animals

OCR B	B2	✓
EDEXCEL	B1	✓
WJEC	B1	✓
CCEA	B1	✓

Humans have been classifying organisms into groups ever since they started studying them:

- This makes it convenient when trying to identify an unknown organism.
- It also tells us something about how closely related organisms are and about their evolution.

The modern system that we use puts organisms into a system of smaller and smaller groups. The groups used are:

- kingdom
- phylum
- class
- order
- family
- genus
- species.

> There are lots of good ways of remembering the order of the groups from kingdom down to species. One example is King Phillip Came Over For Great Spaghetti. You could always make up your own.

> **KEY POINT**
>
> **Kingdoms** are the largest groups. The kingdoms are divided into smaller and smaller groups until the smallest group formed is called a **species**.

As you move down the groups there are fewer organisms in the group and they have more similarities.

Artificial versus natural systems

OCR A	B3	✓
OCR B	B2	✓
EDEXCEL	B1	✓
WJEC	B1	✓
CCEA	B1	✓

> To get an A*, you must realise that organisms can have similar features for two different reasons. They may be closely related or they may be distantly related, but both adapted for living in a similar environment.

The characteristics that are used to classify organisms have changed over time. The system used to be an **artificial system** based on one or two simple characteristics to make identification easier. An example might be the presence of wings on an animal.

Now a **natural system** is used which is based on evolutionary relationships. Animals that are more closely related are more likely to be in the same group. To work out how closely related organisms are it is possible to study their DNA.

The more similar the DNA is, the closer the relationship.

Species

OCR B	B2	✓
EDEXCEL	B1	✓
CCEA	B1	✓

Members of a species are very similar, but how do we know if two similar animals are in the same species?

> **KEY POINT**
>
> Members of the same species can breed with each other to produce fertile offspring.

This means that horses and donkeys are different species because although they can mate and produce a mule, mules are infertile. The mule is an example of a **hybrid**.

Horse + donkey = mule!

Horse   Donkey   Mule

Some organisms cause specific problems when trying to classify them as a species:

- Bacteria do not inter-breed, they reproduce asexually, so they cannot be classified into different species using the 'fertile offspring' idea.
- Hybrids are produced when members of two species inter-breed and so they are infertile. This occurs between many duck species.

Different groups

OCR B	B2	✓
EDEXCEL	B1	✓
WJEC	B1	✓
CCEA	B1	✓

The first step in classifying an organism is to put it into a kingdom. The five kingdoms are shown in the table:

Kingdom	Features
Prokaryotes (bacteria)	No nucleus
Animals	Multicellular, feed on other organisms
Plants	Cellulose cell wall, use light energy to produce food
Protoctista	Mostly single celled with some plant and some animal characteristics
Fungi	Cell wall of chitin, produce spores

Once an organism is put into the animal kingdom it can be put into the **vertebrate** phylum or one of several invertebrate phyla such as the **arthropods**.

> **KEY POINT**
>
> The vertebrates all have a backbone and the group is divided into five different classes:

OCR B candidates also need to know the main characteristics of the arthropod classes (phyla).

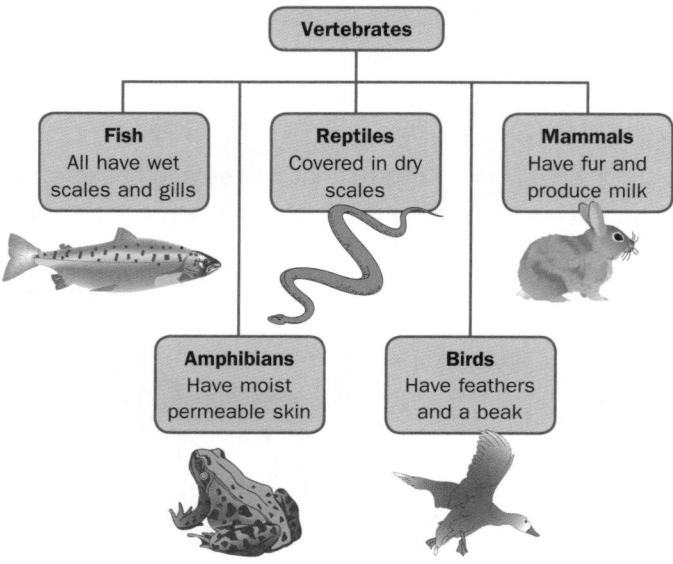

Vertebrates

Fish — All have wet scales and gills

Reptiles — Covered in dry scales

Mammals — Have fur and produce milk

Amphibians — Have moist permeable skin

Birds — Have feathers and a beak

Naming organisms

OCR B	B2	✓
EDEXCEL	B1	✓
WJEC	B1	✓

Organisms are often known by different names in different countries or even in different parts of the same country. All organisms are therefore given a scientific name by the international committees that are used by scientists in every country. This avoids confusion.

> **KEY POINT**
>
> The scientific system of naming organisms is called the **binomial system**.

Each name has two parts. The first part is the name of the genus (the group above species). The second part of the name is the species. For example:

Lion is *Panthera leo*

Tiger is *Panthera tigris*

These animals are in the same genus, but are different species. The genus starts with a capital letter, but the species does not.

PROGRESS CHECK

1. An organism has cell walls made of chitin. What type of organism is it?
2. What is a hybrid?
3. Can tigers and lions mate to produce fertile offspring? Explain your answer.
4. What are the characteristics of mammals?
5. Sharks are fish, but dolphins are mammals. Why are they so similar in appearance?
6. Archaeopteryx is an extinct animal. Fossils show that it had feathers and teeth. Why are there always going to be animals like archaeopteryx that are difficult to classify?

1. A fungus.
2. The offspring of a cross between members of two closely related species.
3. No, because they are not the same species.
4. Have fur and produce milk.
5. They are both adapted to living in the same conditions, i.e. water.
6. Because organisms have evolved from common ancestors by a gradual process there are always going to be organisms that have characteristics that are intermediate between groups.

4.2 Competition and adaptation

LEARNING SUMMARY	**After studying this section, you should be able to:**
	• describe the reasons why organisms compete
	• describe adaptation of organisms to extreme climates
	• describe adaptations of predators and prey.

Competition

AQA	B1	✓
OCR A	B3	✓
OCR B	B2	✓
WJEC	B1	✓
CCEA	B1	✓

There are many different types of organisms living together in a habitat and many of them are after the same things.

KEY POINT

This struggle for resources is called **competition**.

The more similar the organisms, the greater the competition.

Plants usually compete for:

- light for photosynthesis
- water
- minerals.

Organisms of the same species are more likely to compete with each other because they have similar needs.

> **KEY POINT**
>
> A **niche** describes the habitat that an organism lives in and also its role in the habitat.

Organisms that share similar niches are more likely to compete with each other as they require similar resources. There are different types of competition:

- **Intraspecific** is between organisms of the same species and is likely to be more significant as the organisms share more similarities and so need the same resources.
- **Interspecific** is between organisms of different species.

Adaptations to extreme conditions

AQA	B1	✓
OCR A	B3	✓
OCR B	B2	✓
EDEXCEL	B1	✓
WJEC	B1	✓

Because there is constant competition between organisms, the best suited to living in the habitat survive. Over many generations the organisms have become suited to their environment. The features that make organisms well suited to their environment are called **adaptations**. Habitats, such as the Arctic and deserts, are difficult places to live because of the extreme conditions found there. Organisms that are adapted to living in extreme conditions are often called **extremophiles**. Animals and plants have to be well adapted to survive.

Polar bears have:	Cacti have:	Camels have:
A large volume to surface area to minimise heat loss	Leaves that are just spines to reduce surface area to minimise water loss	A hump that stores food as fat
Thick insulating fur	Water stored in the stem	Thick fur on top of the body for shade to protect the skin
A thick layer of fat under the skin		Thin fur on the rest of the body to avoid overheating
White fur that is a poor radiator of heat and provides camouflage		

Be prepared to identify the adaptations on animals that you have not met. Think about size, thickness of fur and body fat.

Adaptation to cold conditions

| AQA | B1 | ✓ |
| OCR B | B2 | ✓ |

To prevent animals losing too much heat in cold climates they are usually quite large, like the polar bear, and have small ears. This helps to decrease the **surface area to volume ratio**. They are more likely to give birth to live young and less likely to lay eggs because the eggs would get too cold before hatching.

All the members of a population may reproduce at the same time, so that predators would not be able to eat all the young. They may try to avoid the coldest temperatures by changing their behaviour. Some animals will **migrate** long distances to warmer areas. Others may stay in the cold areas, but slow down all their body processes and **hibernate**.

When the sun is shining animals like reptiles will lie in the sun or **bask** to try and increase their body temperature.

> **To get an A*, you must be able to explain why similar animals tend to be larger in arctic regions and smaller in desert regions. Remember to talk about surface area to volume ratio.**

Adaptations of predators and prey

| AQA | B1 | ✓ |
| OCR B | B2 | ✓ |

Some animals called **predators** are adapted to hunt other animals for food. The animals that are hunted are called **prey** and are adapted to help them to escape.

Predators are adapted by having:

- Eyes on the front of their head which gives **binocular vision** to judge size and distance.
- Sharp teeth and claws to catch hold of prey.
- A body built for speed to chase prey.
- Stings or **venoms** (poison) to paralyse or poison prey.

Prey animals are adapted by having:

- A body that is **camouflaged** to avoid being seen by predators.
- Eyes on the side of their head to give a view all around.
- A social organisation which involves living in groups which reduces the chance of being caught.
- A body built for speed to outrun predators.
- Defences such as stings or poison to deter predators eating them, along with warning colouration.

Specialists and generalists

| OCR B | B2 | ✓ |

Some organisms, like polar bears, are very well adapted to living in specific habitats. These organisms are called **specialists**. They can survive in these areas when others cannot, but would struggle to live elsewhere.

Other organisms like rats, are not especially adapted to living in one habitat, but can live in many areas. These organisms are called **generalists**. They will be outcompeted by specialists in certain habitats.

PROGRESS CHECK

1. Why do plants compete for light?
2. Why do camels have thick fur on the top of their body and thin fur underneath?
3. What is special about the leaves of cacti?
4. Why is it difficult to creep up behind a rabbit?
5. Emperor penguins lay eggs and stay very close to the South Pole throughout winter. Why does that make them unusual?
6. Elephants need to be large to digest vast quantities of poor quality food. What problem does this large size lead to in the desert and how does the elephant solve this?

1. Light is needed for photosynthesis.
2. Thick fur on top insulates them from the Sun's heat, thin fur underneath allows heat to escape.
3. They are reduced to spines so that less water is lost.
4. They have eyes on the side of their heads so that they have virtually all round vision.
5. Many animals that live near the poles give birth to live young and often migrate away from the pole in the winter. Most animals in cold regions give birth to live young because eggs are likely to get too cold and the foetus may die.
6. Large animals have a small surface area to volume ratio and so could overheat. Elephants have big ears to increase their surface area but could overheat, so they have large ears to increase surface area and lose excess heat.

4.3 Living together

LEARNING SUMMARY	**After studying this section, you should be able to:**
	• explain the shape of predator-prey graphs
	• recall the meaning of the terms parasite and host
	• explain the term mutualism and describe examples.

Predators and prey

OCR B   B2   ✓

Organisms form different types of relationships with other organisms in their habitat. One of the most common is that of predator and prey. The numbers of predators and prey in a habitat will vary and will affect each other. The size of the two populations can be plotted on a graph that is usually called a predator–prey graph.

In the graph the peaks of the predators curve occur a little while after the peaks of the prey curve. This is because it takes a little while for the increase in food supply to allow more predators to survive and reproduce.

A predator–prey graph for lynx and hares.

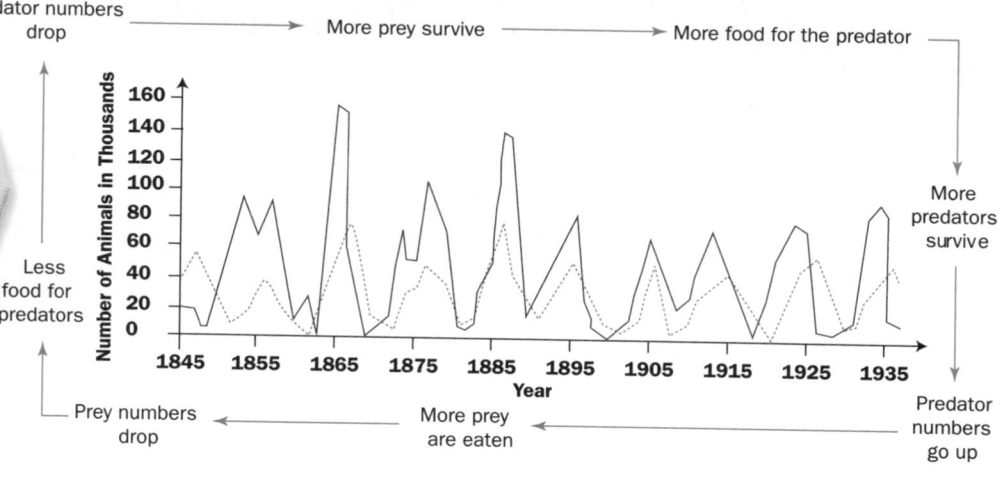

Predator numbers drop → More prey survive → More food for the predator →

More predators survive

Less food for predators

Prey numbers drop ← More prey are eaten ← Predator numbers go up

> Sometimes two different y axes are given on these graphs, one for the predator and the other for the prey. This is because the numbers of predators and prey may be very different. Make sure that you read any figures from the correct scale.

Key

—— Hare ········· Lynx

Parasites and hosts

| OCR B | B2 | ✓ |
| EDEXCEL | B1 | ✓ |

Sometimes one organism may not kill another organism, but it may take food from it while it is alive.

> **KEY POINT**
>
> A **parasite** lives on, or in, another living organism called the **host**, causing it harm.

Fleas are parasites that live in an animal's fur.

Tapeworms are parasites that may live in an animal's gut.

Many diseases, such as **malaria**, are caused by parasites feeding on a host. The parasite in malaria is a single-celled species called Plasmodium that feeds on humans, who are the host. The organism is injected into the bloodstream by a mosquito. This is also acting as a parasite, but is known as a **vector** for malaria because it spreads the disease causing organism, without being affected by it.

Mistletoe is a partial parasite. It grows on trees such as apple trees. It is green so it can photosynthesise and make its own food, but it also takes food from the apple tree.

Mutualism

OCR B	B2	✓
EDEXCEL	B1	✓

Instead of trying to eat each other some different types of organisms work together.

> **KEY POINT**
>
> When two organisms of different species work together so that both gain, it is called **mutualism**.

Examples of this type of relationship are:

- Oxpeckers and buffalo: the oxpecker birds eat the parasites on the backs of the buffalo. So the birds get food and the buffaloes get their parasites removed.
- Cleaner fish: these fish live in certain areas of the reef and are visited by larger fish. They do the same job as oxpeckers.
- Pollinating insects: they visit flowers and so transfer pollen allowing pollination to happen. They are 'rewarded' by sugary nectar from the flower.

> To get an A*, you must be able to link the presence of legumes with their nitrogen fixing bacteria to the nitrogen cycle. This is discussed on page 87.

Pea plants and certain types of bacteria also benefit from mutualism. Pea plants are **legumes** and have structures on their roots called nodules. **Nitrogen fixing bacteria** live in these nodules. The bacteria turn nitrogen gas into nitrogen containing chemicals and give some to the pea plant. The pea plant gives the bacteria some sugars that have been produced by photosynthesis.

Tube worms live deep in the ocean and cannot feed themselves. They have chemosynthetic bacteria living inside them. They can make their own food using the energy from chemical reactions. They give some of this to the worms in return for a safe place to live.

> **PROGRESS CHECK**
>
> 1. When prey numbers are high then predator numbers start to increase. Why is this?
> 2. Why are fleas described as parasites?
> 3. Why is mistletoe called a partial parasite?
> 4. Why do flowers produce nectar?
> 5. Lichens have mutualistic relationships. Algae grow inside the cells of fungi and get water and minerals from the fungi. Suggest what the fungi get in return.
> 6. Why is there so little food available deep in the oceans?
>
> 1. Predators have more food, so are able to reproduce more.
> 2. They live on a living organism and take food from them so cause them harm.
> 3. It can make some of its own food by photosynthesis, but also takes some food from the tree it grows on.
> 4. To attract and reward insects so they pollinate them.
> 5. Fungi cannot photosynthesise so they get food from the algae.
> 6. Very little sunlight can penetrate there so there are no plants as photosynthesis cannot occur.

4.4 Energy flow

LEARNING SUMMARY

After studying this section, you should be able to:

- explain what is meant by a food web
- construct a pyramid of biomass
- calculate the efficiency of energy flow through a food chain.

Food webs

AQA	B1	✓
OCR A	B3	✓
OCR B	B2, B6	✓
EDEXCEL	B1	✓
WJEC	B1	✓
CCEA	B1	✓

A food web shows the feeding relationships between organisms in a habitat.

> **KEY POINT**
>
> Each stage, or feeding level, in a food chain or food web is called a **trophic** level.

Producers are at the start of a food web because they can make their own food. Most producers are green plants or algae that make food by photosynthesis. Very few are bacteria, such as the ones that live in tube worms, who make food using energy from chemical reactions (chemosynthesis). Very few organisms only eat one type of food. Most will eat several types and the food might be from different trophic levels. For example in this food web the birds eat both ladybirds and blackfly. When they eat blackfly they are secondary consumers and when they eat ladybirds they are tertiary consumers.

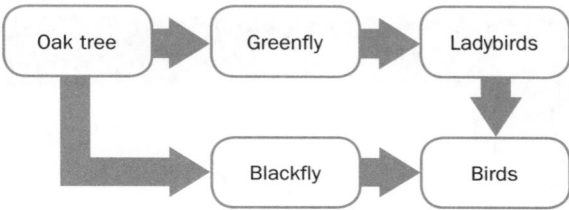

If one organism is reduced or increases in numbers in a food web it can alter the numbers of other organisms in the food web.

Pyramids of biomass

AQA	B1	✓
OCR B	B2	✓
EDEXCEL	B1	✓
WJEC	B1	✓
CCEA	B1	✓

The mass of all the organisms at each step of the food chain can be estimated. This can be used to draw a diagram that is similar to a pyramid of numbers. The difference is that the area of each box represents the mass of all the organisms not the number. This type of diagram is called a **pyramid of biomass**.

The reason that a pyramid of biomass is shaped like a pyramid is that energy is lost from the food chain in different ways as the food is passed along. Often the waste from one food chain can be used by decomposers to start another chain.

Where energy is wasted in the food chain.

Reflection | Waste and uneaten parts | Waste, uneaten parts, and heat through respiration

> **Remember excretion is the removal of waste products made by the body, e.g. urine, whereas egestion is food material that passes straight through. Candidates often get these two confused.**

The diagram shows that biomass and energy are lost from the food chain in a number of ways:

- In waste from the organisms by **excretion** and **egestion**.
- As heat when organisms **respire**. Birds and mammals that keep a constant body temperature will often lose large amounts of energy in this way.

Measuring biomass

OCR B	B2	✓
CCEA	B1	✓
WJEC	B1	✓

Although pyramids of biomass are a better way of representing trophic levels they are difficult to construct. This is because:

- Some organisms feed on organisms from different trophic levels.
- Measuring dry mass is difficult as it involves removing all the water from an organism which will kill it.

Calculating efficiency

AQA	B1	✓
OCR A	B3	✓
OCR B	B2	✓
EDEXCEL	B1	✓
WJEC	B1	✓
CCEA	B1	✓

The diagram shows the flow of energy through a food chain.

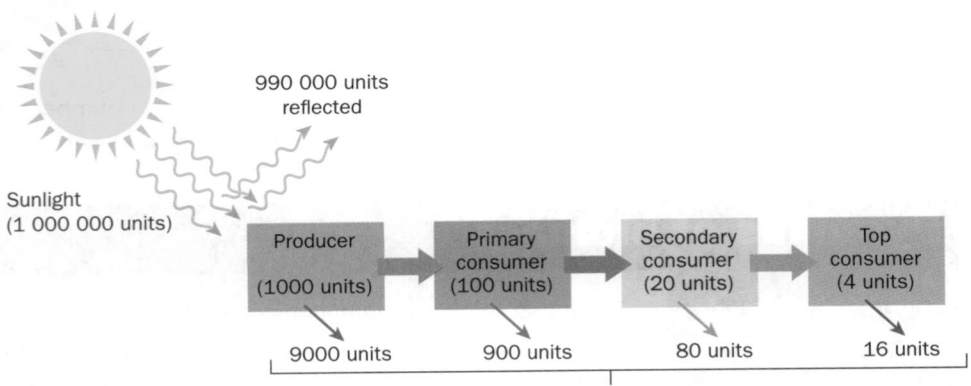

Sunlight (1 000 000 units) — 990 000 units reflected

| Producer (1000 units) | Primary consumer (100 units) | Secondary consumer (20 units) | Top consumer (4 units) |

9000 units | 900 units | 80 units | 16 units

All lost from food chain

> **To get an A*, you must be able to work out the percentage efficiency of energy transfer in different food chains and suggest reasons for differences. For example, farmers often keep their cattle indoors so that they lose less heat keeping warm.**

Of one million units of light that hit the surface of the plant, only 4 units are used for growth in the top consumer. This loss of energy also explains why food chains usually only have four or five steps. By then, there is so little energy left the animals would not be able to find enough food.

The percentage efficiency of transfer from producer to primary consumer in the diagram is:

$$\frac{100}{1000} \times 100\% = 10\%$$

This percentage is quite low because it is difficult to digest plant material.

PROGRESS CHECK

1. Why are food webs more common than simple food chains in nature?
2. Look at the food web on page 83. What might happen to the numbers of ladybirds if a gardener killed all the blackfly in his garden with insecticide?
3. What does the size of a box represent in a pyramid of biomass?
4. Write down two ways that energy is lost from a food web.
5. Calculate the percentage efficiency of the energy transfer from secondary to top consumer in the above food chain.
6. Compare your answer from question 5 with the percentage efficiency from producer to primary consumer. Suggest reasons for the difference.

6. This is much more efficient. This is because meat contains more energy and it is easier to digest.
5. $\frac{4}{20} \times 100\% = 20\%$
4. Heat/excretion/egestion/uneaten parts from respiration.
3. The amount of living material at each trophic level.
2. They might decrease because the birds would eat more of them as there are no blackfly to eat.
1. Most organisms eat more than one type of food.

4.5 Recycling

LEARNING SUMMARY

After studying this section, you should be able to:

- describe the conditions needed by decomposers
- interpret diagrams of the carbon and nitrogen cycles.

Decay

AQA	B1	✓
OCR A	B3	✓
OCR B	B2, B4,	✓
	B6	✓
WJEC	B1	✓
CCEA	B1	✓

Some animals and plants die before they are eaten. They also produce large amounts of waste products. This waste material must be broken down or **decayed** because it contains useful minerals. If this did not happen, organisms would run out of minerals. Ecosystems are therefore called **closed loop systems** as the minerals are constantly recycled and not lost.

KEY POINT

Organisms that break down dead organic material are called **decomposers**.

The main organisms that act as decomposers are bacteria and fungi.

They release enzymes on to the dead material that then digest the large molecules. They then take up the soluble chemicals that are produced. The bacteria and fungi use the chemicals in respiration and for raw materials. For decomposers to decay dead material they need certain conditions:

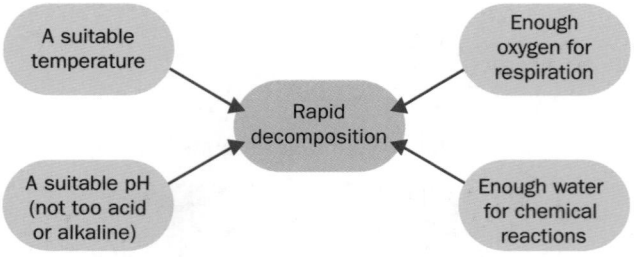

> Gardeners try to produce ideal conditions for decay in their compost heaps. Make sure you can explain how they do this.

Earthworms are important for decomposition in the soil. This is because they drag dead leaves below the surface and also aerate the soil.

The carbon cycle

AQA	B1	✓
OCR A	B3	✓
OCR B	B2	✓
EDEXCEL	B1	✓
WJEC	B1	✓
CCEA	B1	✓

It is possible to follow the way in which each mineral element passes through living organisms and becomes available again for use. Scientists use nutrient cycles to show how these minerals are recycled in nature. One of these is the carbon cycle.

Carbon dioxide is returned to the air in a number of different ways:

- Plants and animals respire.
- Soil bacteria and fungi acting as decomposers.
- The burning of fossil fuels (combustion).

The main process that removes carbon dioxide from the air is photosynthesis.

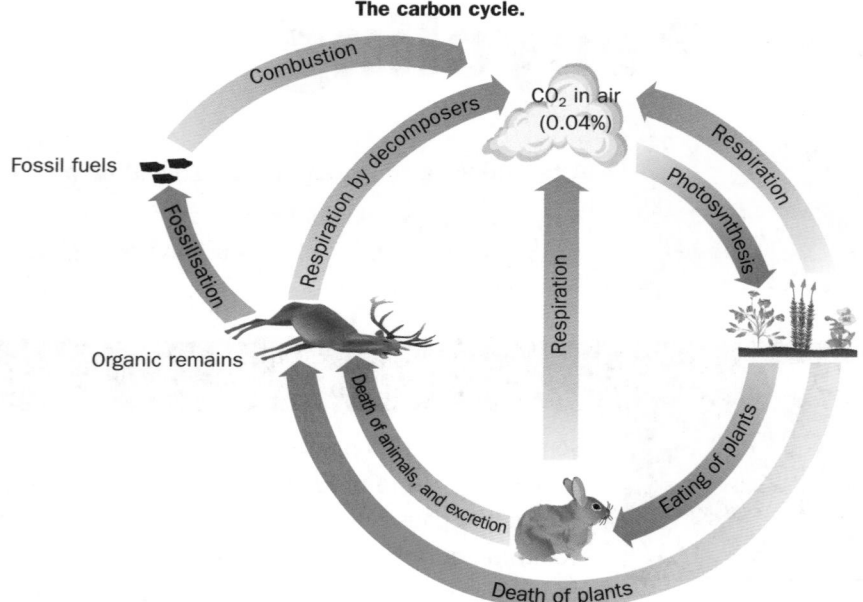

The carbon cycle.

The carbon cycle in the sea

| OCR B | B2 | ✓ |

Carbon dioxide is absorbed from the air by oceans. Much of this is by algae. Some marine organisms use the carbon dioxide and make shells made of carbonate which, over millions of years, become limestone rocks.

The carbon in limestone can return to the air as carbon dioxide during volcanic eruptions or weathering. The action of acid rain on buildings will speed up the weathering process due to the reaction with the limestone.

The nitrogen cycle

OCR A	B3	✓
OCR B	B2	✓
EDEXCEL	B1	✓
WJEC	B1	✓
CCEA	B1	✓

Plants take in nitrogen as **nitrates** from the soil to make protein for growth. Feeding passes nitrogen compounds along a food chain or web. The nitrogen compounds in dead plants and animals are broken down by decomposers and returned to the soil.

The nitrogen cycle is more complicated than the carbon cycle because as well as the decomposers, it involves three other types of bacteria:

- **Nitrifying bacteria** – these bacteria live in the soil and convert ammonium compounds to nitrates. They need oxygen to do this.
- **Denitrifying bacteria** – these bacteria in the soil are the enemy of farmers. They turn nitrates into nitrogen gas. They need conditions without oxygen, rather than needing oxygen.
- **Nitrogen fixing bacteria** – they live in the soil or in special bumps called nodules on the roots of plants from the pea and bean family. They take nitrogen gas and convert it back to useful nitrogen compounds.

The nitrogen cycle.

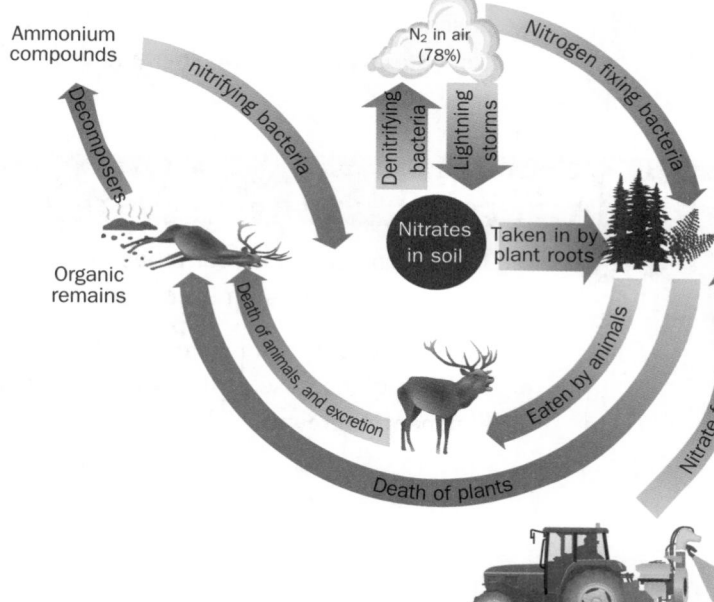

To get an A*, you must be able to recognise parts of the nitrogen cycle. You are unlikely to have to describe it all, but you might, for example, have to explain how the protein in fallen leaves gets converted to nitrates.

Although the air contains about 78% nitrogen it is unreactive. This is why lightning and nitrogen fixing bacteria are so important. They **fix** the nitrogen back into chemicals that can be used by plants.

PROGRESS CHECK

1. Gardeners water their compost heaps in dry weather. Why do they do this?
2. Write down two ways that carbon dioxide is returned to the atmosphere.
3. What is the main nitrogen containing compound taken up by plants?
4. What do plants need nitrogen for?
5. Explain why acid rain releases carbon dioxide when it falls onto certain buildings.
6. Farmers try to make sure that their soils are well drained so that they contain plenty of air. Explain why they do this.

(cont.)

4.6 Pollution and overexploitation

<table>
<tr><td rowspan="3">LEARNING
SUMMARY</td><td>After studying this section, you should be able to:</td></tr>
<tr><td>• describe problems associated with an increase in the human population
• describe sources and effects of different pollutants
• compare indicator organisms and direct methods of measuring pollution.</td></tr>
</table>

Population increase

AQA	B3	✓
OCR B	B2	✓
EDEXCEL	B1	✓

The human population on Earth has been increasing for a long time, but it is now going up more rapidly than ever:

- This is because of an increasing birth rate and decreasing death rate.
- The rate of increase of the population is increasing and this is called **exponential growth**.

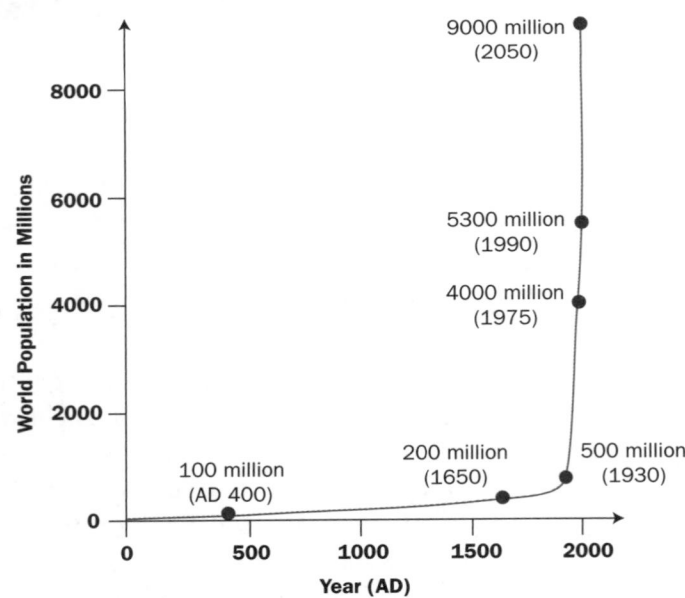

Population growth.

This increase in the population is having a number of effects on the environment:

- More raw materials are being used up such as fossil fuels and minerals.
- More waste is being produced which can lead to pollution.
- More land is being taken up to be used for activities such as building, quarrying, farming and dumping waste.

Pollution

AQA	B3	✓
OCR B	B2	✓
EDEXCEL	B1	✓
WJEC	B1	✓
CCEA	B1	✓

Modern methods of food production and the increasing demand for energy have caused many different types of **pollution**.

> **KEY POINT**
>
> Pollution is the release of substances into the environment that harm organisms.

The table shows some of the main polluting substances that are being released into the environment.

Polluting substance	Main source	Effects on the environment
Carbon dioxide	Burning fossil fuels	Greenhouse effect
Sulfur dioxide	Burning fossil fuels	Acid rain
Chlorofluorocarbon (CFCs)	Fridges and aerosols	Destroys the ozone layer
Fertilisers including nitrates and phosphates	Intensive farming	Pollutes rivers and lakes
Domestic waste	Households	Landfill sites release gases
Heavy metals	Industry	Accumulates in food chains
Sewage	Human and farm waste	Pollutes rivers and lakes

The **greenhouse effect** is caused by a build-up of gases, such as carbon dioxide and methane, in the atmosphere. These gases trap the heat rays as they are radiated from the Earth. This causes the Earth to warm up. This is similar to what happens in a greenhouse. This could lead to changes in the Earth's climate and a large rise in sea level.

The greenhouse effect.

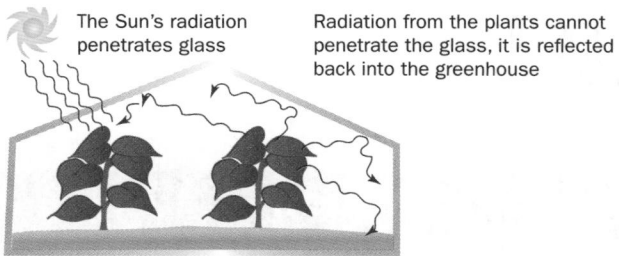

The Sun's radiation penetrates glass

Radiation from the plants cannot penetrate the glass, it is reflected back into the greenhouse

Acid rain is caused by the burning of fossil fuels that contain sulfur impurities. These give off sulfur dioxide, which dissolves in rainwater to form sulfuric acid. This falls as acid rain.

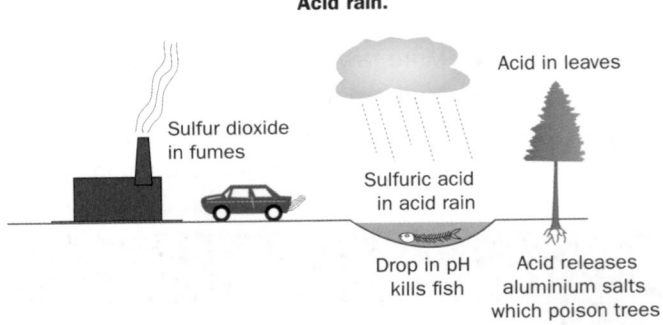

Acid rain.

Sulfur dioxide in fumes

Acid in leaves

Sulfuric acid in acid rain

Drop in pH kills fish

Acid releases aluminium salts which poison trees

> Remember that CFCs are greenhouse gases as well as breaking down the ozone layer, but it is the lack of the ozone layer that can cause skin cancer.

Ozone depletion is caused by the release of chemicals such as CFCs which come from the breakdown of refrigerators and aerosol sprays. Ozone helps protect us from harmful ultraviolet (UV) radiation and so depletion may lead to more skin cancer.

Carbon footprints

| OCR B | B2 | ✓ |

The world population figures show that the greatest rise in population is occurring in countries such as Africa and India. However, the countries that use the most fossil fuels are developed countries, such as the USA and Europe.

A useful way of measuring how much pollution is caused per person is the **carbon footprint**. This measures the total greenhouse gas given off by a person, or organisation, over a certain period.

Eutrophication

OCR A	B7	✓
OCR B	B6	✓
EDEXCEL	B1	✓
WJEC	B1	✓
CCEA	B1	✓

Rain water containing fertilisers can run-off from fields into rivers. Similarly, sewage also pollutes rivers. This can result in **eutrophication** of the river or lake:

- Nitrates and phosphates enter rivers and are absorbed by plants and algae. This makes them grow.
- Algae float near the water surface and their population increases dramatically. A 'blanket' of algae soon covers the surface.
- Other plants beneath the algae die as the surface algae block out the sunlight.
- Bacteria and other decomposers begin to break down the dead plants using more oxygen for respiration. Fish die as the oxygen content of the water becomes too low.

> To get an A*, you must be able to link the nitrogen cycle to eutrophication. Remember that nitrates in fertilisers are needed by plants to produce protein and are also produced when sewage is decomposed to ammonium compounds and then converted to nitrates by nitrifying bacteria.

Indicators of pollution

AQA	B1	✓
OCR B	B2	✓
EDEXCEL	B1	✓
WJEC	B1	✓
CCEA	B1	✓

Some organisms are more sensitive to pollution than others. If we look for these organisms it can tell us how polluted an area is. On land, lichens grow on trees and stone. Some lichens are killed by lower levels of pollution than other types. Black spot fungus grows on roses in areas with less sulfur dioxide pollution.

In water some animals, for example rat-tailed maggots, can live in polluted water, but other animals like mayfly larvae can only live in clean water.

There are advantages to the different methods of measuring pollution:

- Using **indicator organisms** is cheaper, does not need equipment that can go wrong and monitors pollution levels over long periods of time.
- Using **direct methods** can give more accurate results at any specific time.

PROGRESS CHECK

1. The world's population is showing an exponential increase. What does this mean?
2. Write down **two** gases that can cause the greenhouse effect.
3. What is the main gas that causes acid rain?
4. Write down one indicator organism that is found in polluted water.
5. Explain why eating food that is grown in another country will result in a higher carbon footprint.
6. Nitrate fertilisers cause eutrophication. Explain how sewage can also cause eutrophication.

1. The rate of increase is increasing.
2. Carbon dioxide/methane.
3. Sulfur dioxide.
4. Rat-tailed maggots.
5. Transportation of that food to us involves the burning of fossil fuels that release carbon dioxide.
6. Sewage contains nitrogen containing compounds, so can also cause an algal bloom as the algae use it for growth.

4.7 Conservation and sustainability

	After studying this section, you should be able to:
LEARNING SUMMARY	• explain why some organisms are at risk of extinction • discuss reasons for maintaining biodiversity • describe the principles of conservation programmes • explain the importance of maintaining genetic variation • explain what is meant by sustainable development.

Organisms at risk

AQA	B2, B3	✓
OCR A	B7	✓
OCR B	B2	✓
WJEC	B2	✓
CCEA	B1	✓

As well as causing pollution, the increasing demands for food, land and timber have caused people to cut down large areas of forests. Deforestation has led to:

- Less carbon dioxide being removed from the air by the trees and carbon dioxide being released when the wood is burnt.
- The destruction of habitats which contain rare species.

Some animals have been hunted, so their numbers have been dramatically reduced, for example, species of whales which have been hunted for food, oil and other substances. Their numbers now are very low and people are trying to protect them. Other organisms have not been so lucky. Their numbers have decreased so far that they have completely died out. This is called **extinction**. Organisms do become extinct naturally, but man has often increased the rate either directly or indirectly by:

- over-hunting
- destroying habitats
- pollution
- competition
- changing the climate.

Other organisms are at risk of becoming extinct and are **endangered**.

Biodiversity

AQA	B3	✓
OCR A	B3, B7	✓
OCR B	B2, B4	✓
WJEC	B2	✓
CCEA	B1	✓

Many people believe that it is wrong for humans to damage natural habitats and cause the death of animals and plants. They believe that it is important to keep a wide variety of different animals and plants alive.

> **KEY POINT**
>
> The variety of different organisms that are living is called **biodiversity**.

There are many reasons given for trying to maintain biodiversity, such as:

- Losing organisms may have unexpected effects on the environment, such as the erosion caused by deforestation.
- Losing organisms may have effects on other organisms in their food web.
- Some organisms may prove to be useful in the future, such as for breeding, producing drugs or for their genes.
- Organisms may be needed for food.
- People enjoy looking at and studying different organisms.

Conservation programmes

OCR A	B3, B7	✓
OCR B	B2, B4	✓
WJEC	B2	✓
CCEA	B1	✓

To help save and preserve habitats and organisms, people have set up many different **conservation** schemes. There are a number of different ways that conservation programmes can work:

> Questions on this part of the course could use an animal or plant that you have never heard of. You may need to use the data that you are provided with to try and work out why it became endangered and what could be done to try and protect it.

How to save an endangered species.

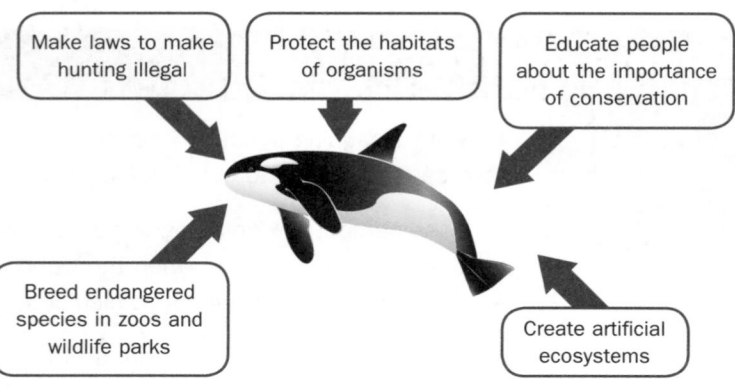

Make laws to make hunting illegal

Protect the habitats of organisms

Educate people about the importance of conservation

Breed endangered species in zoos and wildlife parks

Create artificial ecosystems

Genetic variation

OCR B   B2   ✓

If populations become small then the variety of different alleles in the population might be quite low. The population has low **genetic variation**. This makes it more likely to become extinct because:

- It will be harder for it to adapt to any changes in the environment because there is little variation.
- There is more chance of organisms being produced that have two identical harmful recessive alleles.

KEY POINT

Reproducing with an organism that has similar alleles is called **inbreeding**.

Zoos try and move animals to other zoos to breed with less related organisms to avoid this happening. It is also more likely to happen when populations become isolated.

Sustainable development

AQA	B3	✓
OCR A	B3, B7	✓
OCR B	B2	✓
WJEC	B1	✓

If the human population is going to continue to increase, it is important that we meet the demand for food and energy without causing pollution or over-exploitation.

KEY POINT

Providing for the increasing population without using up resources or causing pollution is called **sustainable development**.

Fish stocks and woodland can be managed sustainably by:

- Educating people about the importance of controlling what they take from the environment.
- Putting quotas on fishing.
- Re-planting woodland when trees have been removed.

A decrease in the use of packaging materials and recycling would also help by:

- Cutting down the energy needed to make them and transport them.
- Reducing the problem of disposing of the waste.

Alternatives to peat composts must also be found. This will help to prevent the destruction of peat bogs. These are rare habitats and the removal and decomposition of the peat adds carbon dioxide to the air.

To make sure that development is sustainable a lot of planning is needed at local, national and international levels.

PROGRESS CHECK

1. Write down two ways that man has caused extinctions.
2. What is biodiversity?
3. What is conservation?
4. Why does recycling help to improve sustainability?
5. Areas of tropical rainforest are being cleared, but small patches are being left. Explain why small isolated areas may not be very useful in conserving organisms.
6. About 10 000 years ago the cheetah nearly became extinct. Zoos find it very difficult to get cheetahs to produce healthy offspring. To try and make sure this happens they try and artificially inseminate them using sperm from around the world. Explain these observations.

1. Overhunting/destroying habitats/causing pollution/changing of climate.
2. The variety of different organisms in an environment.
3. Trying to preserve a habitat so that the organisms that live there are protected.
4. Fewer resources are used to make new materials and there is less waste.
5. The small patches may only contain small numbers of individuals of a species. There might not be enough genetic diversity.
6. Cheetahs do not have much genetic diversity as they are all related to a small number of individuals. They try and reproduce them with other cheetahs that are not closely related and so have fewer similarities in their genes. Using artificial insemination from less closely related individuals will help reduce inbreeding problems.

Sample GCSE questions

1 The graph shows how the levels of carbon dioxide have changed in the atmosphere over the past thousand years.

It also shows an estimate of the global temperature.

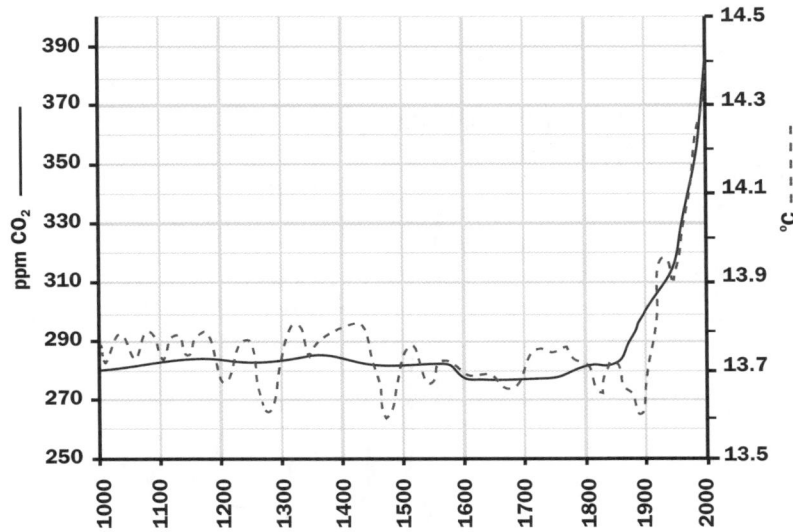

(a) Some scientists say that there is a correlation between carbon dioxide levels and the temperature.

What does this mean? **[2]**

This means that as the carbon dioxide levels go up so does the global temperature.

(b) Use ideas about the greenhouse effect to explain a possible cause of this correlation. **[3]**

Increasing levels of carbon dioxide cause the greenhouse effect.

The gas allows the Sun's radiation through to the Earth but prevents it being re-radiated back out into space. This causes the global temperature to increase.

(c) An alternative theory says the Earth warmed up naturally, allowing more animals to survive.

How could this explain the correlation? **[3]**

More animals survive and so there is more respiration.

This produces carbon dioxide and so levels in the air will increase.

It is important that you know the difference between a correlation and a cause. This answer describes a correlation, it does not say that one factor causes the other.

This is a good summary of the cause of the greenhouse effect but you could say that short wavelength radiation can enter but long wavelength cannot escape.

A good answer. Some candidates may just say that respiration increases and not link it to increased carbon dioxide. Others may just say that there is more carbon dioxide given out.

Exam practice questions

1 The diagram shows part of the nitrogen cycle.

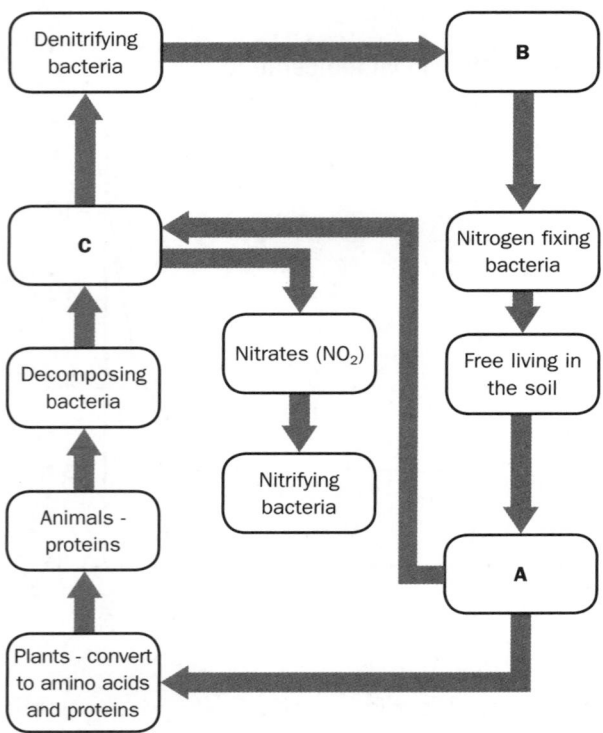

(a) The letters **A**, **B** and **C** represent three chemicals that contain nitrogen.

Write down the names of these three chemicals.

A =_____

B =_____

C =_____   [3]

(b) **(i)** The cycle shows bacteria decomposing dead material.

Name one other type of organism that decomposes dead material.

....

.... [1]

(ii) For decomposers to break down dead material they need certain conditions.

Write down three of these conditions.

....

....

.... [3]

Exam practice questions

(c) The nitrogen fixing bacteria shown on the diagram live free in the soil.

Others live in plants such as peas and beans.

Write about where in the plant they live and about the relationship that they have with the plant.

The quality of written communication will be assessed in your answer to this question.

....

....

....

.... **[5]**

2 The camel has the scientific name *Camelus dromedarius*.

It lives mainly in North Africa.

(a) **(i)** Finish this table to show how the camel is classified.

Kingdom	
Phylum	
Class	
Order	Cetartiodactyla
Family	Camelidae
Genus	
Species	

[5]

(ii) Why are organisms given scientific names?

....

.... **[2]**

Exam practice questions

(b) One of the main habitats for camels is the desert.

Explain how the following special adaptations help the camel survive in the desert:

(i) Webs of skin between their toes.

....

.... **[1]**

(ii) A store of fat in a hump which is on the top of their body.

....

....

.... **[2]**

3 The diagram shows all the energy taken in and lost by a cow in a certain amount of time.

925 kJ as heat

2000 kJ in all waste materials

3050 kJ in cereal

(a) Work out the amount of energy trapped in new tissue in the cow in this time. **[1]**

answer = ... kJ

(b) Calculate the percentage efficiency of energy transfer between the cereal and the cow.

answer = ... %

(c) In terms of energy capture it is more efficient for a person to eat cereal, rather than

feed the cereal to the cow and eat beef.

Explain why.

....

.... **[2]**

5 Cells and molecules

The following topics are covered in this chapter:

- Cells and organisation
- DNA and protein synthesis
- Proteins and enzymes
- Cell division
- Growth and development
- Transport in cells
- Respiration

5.1 Cells and organisation

After studying this section you should be able to:

- describe the differences between plant and animal cells
- describe the fine detail of cells that can be seen with the electron microscope
- describe the levels of organisation found in multicellular organisms
- describe the structure of a typical bacterial cell.

Plant and animal cells

AQA	B2	✓
OCR A	B4	✓
OCR B	B3	✓
EDEXCEL	B2	✓
WJEC	B2	✓
CCEA	B1	✓

Plants and animal cells have a number of structures in common. They all have:

- A **nucleus** that carries genetic information and controls the cell.
- A **cell membrane** which controls the movement of substances in and out of the cell.
- **Cytoplasm** where most of the chemical reactions happen.

There are three main differences between plant and animal cells:

- Plant cells have a strong **cell wall** made of cellulose, whereas animal cells do not. The cell wall supports the cell and stops it bursting.
- Plant cells have a large permanent **vacuole** containing cell sap, but vacuoles in animal cells are small and temporary. The cell sap is under pressure and this supports the plant.

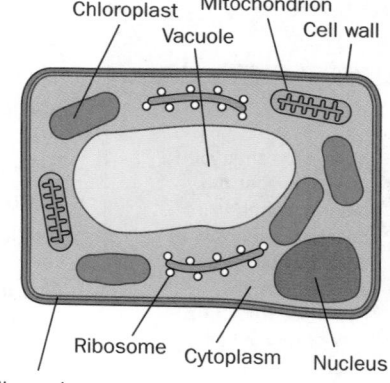

Typical plant cell.

Chloroplast Mitochondrion Vacuole Cell wall Ribosome Cytoplasm Nucleus Cell membrane

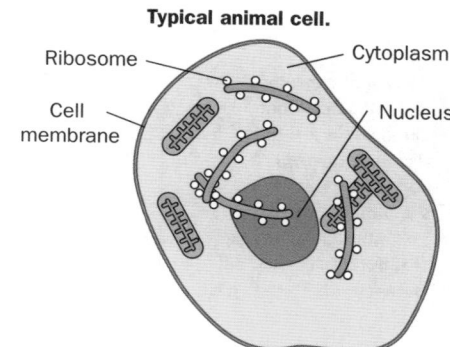

Typical animal cell.

Ribosome Cytoplasm Cell membrane Nucleus

To get an A*, you must be able to measure cells and structures from a drawing and then use the magnification of the drawing to work out their size in real life. Remember that magnification = image size/size in real life.

- Plant cells may contain **chloroplasts** containing chlorophyll for photosynthesis. Animal cells never contain chloroplasts.

The naked eye can see detail down to about 0.1 mm. Cells are smaller than this so a microscope is needed to see individual cells. The best **light microscopes** can magnify cells so that objects as small as 0.002 mm can be seen clearly. At this magnification other structures in the cell become visible, but cannot be seen clearly. Using an **electron microscope** allows objects as small as 0.000002 mm to be seen:

- **Mitochondria** are the site of respiration in the cell.
- **Ribosomes** are small structures in the cytoplasm where proteins are made.

A mitochondrium.

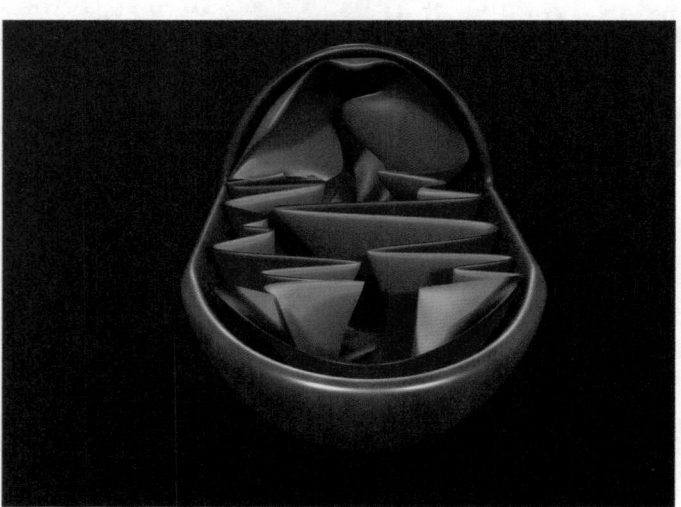

Levels of organisation

AQA	B2	✓
OCR A	B5	✓
OCR B	B3	✓
EDEXCEL	B2	✓

Some organisms are made of one cell. They are **unicellular**. There seems to be a limit in the size of a single cell so larger organisms are made up of a number of cells. They are **multicellular**. The cells are not all alike, but are specialised for particular jobs, for example guard cells in leaves, and neurones.

Similar cells, doing similar jobs, are gathered together into **tissues** such as xylem and nerve. Different tissues are gathered together into **organs** to do a particular job, for example leaves and the brain.

Groups of organs often work together in **systems** to carry out related functions.

Make sure that you can remember the order of terms to describe the levels of organisation: Cells, Tissues, Organs, Systems. Think of a way of remembering this.

Be careful with 'bone' and 'muscle'. Bone and muscle are both tissues. However 'a bone' and 'a muscle' are both organs as they contain a number of different tissues.

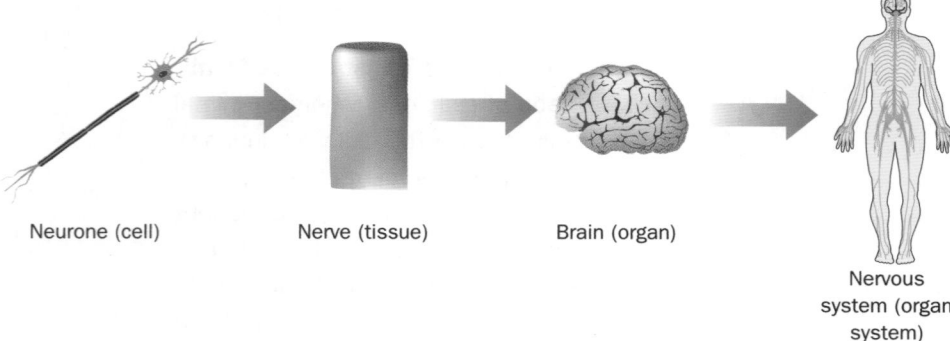

Neurone (cell)　　　Nerve (tissue)　　　Brain (organ)　　　Nervous system (organ system)

Challenges for multicellular organisms

AQA	B2	✓
OCR B	B3	✓
CCEA	B1	✓
WJEC	B2	✓

Becoming larger and multicellular does have some advantages. For example, it allows cells to specialise. This makes them more efficient at their job. However, it also produces difficulties.

The difficulties that need to be solved are:

- It requires a communication system between cells to be developed.
- It is harder to supply all the cells with nutrients.
- The surface area to volume ratio is smaller so it is harder to exchange substances with the environment.

Structure of bacteria

AQA	B2	✓
OCR A	B4	✓
OCR B	B3, B6	✓
EDEXCEL	B2	✓
WJEC	B2	✓
CCEA	B1	✓

Bacterial cells are very different to plant and animal cells. That is why they are classified in a different kingdom (prokaryotes). They vary in shape, but all bacterial cells have a similar structure.

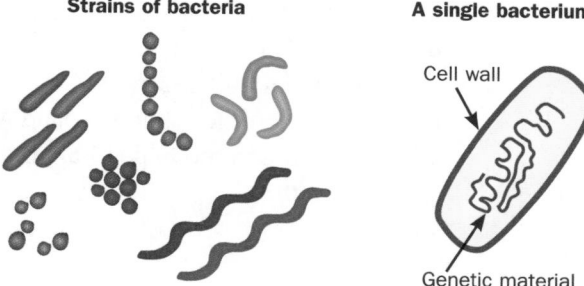

A generalised bacterium.

Strains of bacteria

A single bacterium

Cell wall

Genetic material

Bacterial cells are smaller than animal and plant cells. They lack a true nucleus, mitochondria, chloroplasts and vacuoles.

PROGRESS CHECK

1. Write down three structures that are found in plant cells, but not in animal cells.
2. Why is cell sap important to the structure of a plant?
3. At what level of organisation are leaves and the brain?
4. Where is DNA found in a bacterium?
5. Specialisation allows cells to become more efficient at their particular role. A disadvantage of specialisation is that an organism cannot live if it loses certain organs. Suggest why this is.

5. The cells become so specialised that they cannot take on other roles and take the function of those lost.
4. In the cytoplasm.
3. Organs.
2. Cell sap in the vacuole is under pressure and so supports the plant.
1. Cell wall, vacuole, chloroplasts.

5.2 DNA and protein synthesis

LEARNING SUMMARY

After studying this section you should be able to:

- describe the structure of DNA and how it was discovered
- explain how DNA codes for proteins
- describe the process of protein synthesis.

The structure of DNA

AQA	B2	✓
OCR A	B5	✓
OCR B	B3	✓
EDEXCEL	B2	✓
WJEC	B2	✓
CCEA	B2	✓

The nucleus controls the chemical reactions occurring in the cell. This is because it contains the genetic material. This is contained in structures called **chromosomes** which are made of **DNA**. DNA is a large molecule with a very important structure:

- It has two strands.
- The strands are twisted to make a shape called a **double helix**.
- Each strand is made of a long chain of molecules including sugar, phosphates and bases.
- There are only four bases called A, C, G and T.
- Links between the bases hold the two chains together, C always links with G and A with T.

The DNA helix.

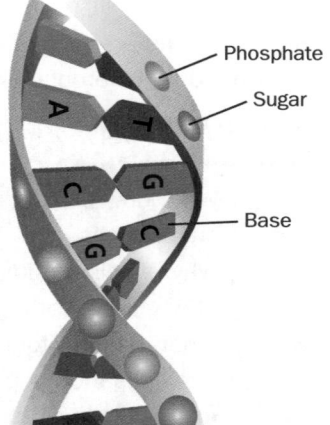

Phosphate
Sugar
Base

> You need to remember **A** with **T** and **C** with **G**. Find a way of remembering it that means something to you. It could be **Auntie Tina** and **Cousin George** for example.

Discovering the structure of DNA

OCR B	B3	✓
EDEXCEL	B2	✓
CCEA	B2	✓

KEY POINT

The two scientists that are famous for discovering the structure of DNA are Francis Crick and James Watson.

'How Science Works' questions may ask for examples of how advances in science are made by cooperation between scientists. The discovery of the structure of DNA is a good example to use.

They worked together in Cambridge in the early 1950s. A molecule of DNA is only about 0.00000034 mm wide, so they could not use a microscope to see it! This is where they needed information obtained by other scientists:

- Maurice Wilkins and Rosalind Franklin fired X-rays at DNA crystals and the images they obtained told Watson and Crick that DNA was shaped like a helix, with two chains.
- Erwin Chargaff had worked out that there was always the same percentage of the base C as G and the same percentage of A as T.

These two pieces of information allowed Watson and Crick to build their famous model of the structure.

Coding for proteins

AQA	B2	✓
OCR A	B5	✓
OCR B	B3	✓
EDEXCEL	B2	✓
WJEC	B2	✓
CCEA	B2	✓

KEY POINT

DNA controls the cell by carrying the code for proteins.

- Each different protein is made of a particular order of amino acids, so DNA must code for this order.
- A gene is a length of DNA that codes for the order of amino acids in one protein.

Scientists now know that each amino acid in a protein is coded for by each set of three bases along the DNA molecule.

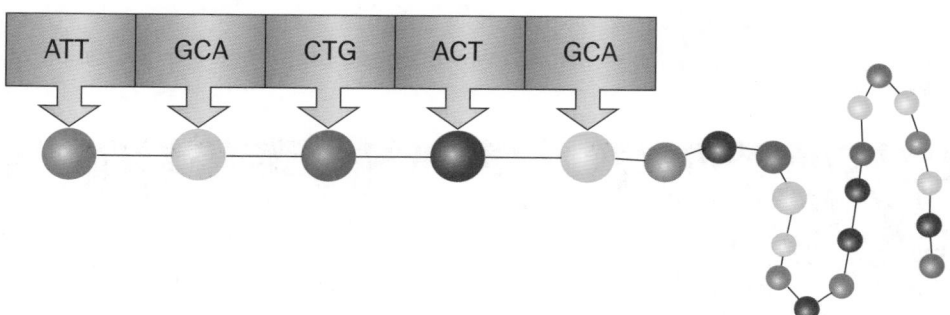

ATT   GCA   CTG   ACT   GCA

Protein synthesis

AQA	B2	✓
OCR A	B5	✓
OCR B	B3	✓
EDEXCEL	B2	✓
WJEC	B2	✓

KEY POINT

Proteins are made on **ribosomes** in the cytoplasm and DNA is kept in the nucleus and cannot leave.

The cell has to use a messenger molecule to copy the message from DNA and to carry the code to the ribosomes. This molecule is called **messenger RNA (mRNA)**. When a protein is to be made these steps occur:

The word complementary is a useful word to use when describing mRNA. It does not 'copy' the code exactly, but makes a version using the matching bases.

- The DNA containing that gene unwinds and 'unzips'.
- Complementary mRNA molecules pair up next to the DNA bases on one strand.
- The mRNA units join together and make a molecule with a complementary copy of the gene. This is called **transcription**.
- The mRNA molecule then leaves the nucleus and attaches to a ribosome.
- The base code on the mRNA is then used to link amino acids together in the correct order to produce the protein. Each three bases code for one amino acid. This is called **translation**.

PROGRESS CHECK

1. Why is the shape of DNA described as a double helix?
2. How does the structure of one protein differ from the structure of another protein?
3. What holds the two strands together in a DNA molecule?
4. Why is the genetic code described as a triplet code?
5. In a length of DNA 34% of the bases are the base G. What percentage are base T?
6. When a protein is to be made the length of DNA containing that gene 'unzips'. What does this mean and why is it necessary?

1. Two strands twisted into a spiral shape.
2. It has a different order of amino acids.
3. Hydrogen bonds between the bases, C with G and A with T.
4. Each three bases code for one amino acid.
5. 16%
6. The double helix unwinds. This allows mRNA bases to pair with the DNA bases producing a complementary copy to pass to the ribosomes.

5.3 Proteins and enzymes

LEARNING SUMMARY

After studying this section you should be able to:

- describe the main functions of proteins
- explain how enzymes work
- explain why enzymes are affected by extremes of temperature and pH.

The functions of proteins

AQA	B2	✓
OCR A	B1	✓
OCR B	B3, B1	✓
WJEC	B2	✓

The only way that the genetic material can control the cell is by coding for which proteins are made. The proteins that are produced have a wide range of different functions:

- Structural proteins used to build cells, e.g. collagen.
- Hormones to carry messages, e.g. insulin.
- Carrier molecules, e.g. haemoglobin.
- Enzymes to speed up reactions, e.g. amylase.

Enzymes

AQA	B2	✓
OCR A	B4	✓
OCR B	B3	✓
EDEXCEL	B2	✓
WJEC	B2	✓
CCEA	B1	✓

KEY POINT

Enzymes are **biological catalysts**.

Many people think that all enzymes are released into the gut to digest food. Remember that most enzymes are found inside cells and are never released.

They are produced in all living organisms and control all the chemical reactions that occur. Most of the chemical reactions that occur in living organisms would occur too slowly without enzymes. Increased temperatures would speed up the reactions, but using enzymes means that the reactions are fast enough at 37°C. These reactions include DNA replication, digestion, photosynthesis, respiration and protein synthesis.

How do enzymes work?

AQA	B2	✓
OCR A	B4	✓
OCR B	B3	✓
EDEXCEL	B2	✓
WJEC	B2	✓
CCEA	B1	✓

As enzymes are protein molecules they are made of a long chain of amino acids that is folded up to make a particular shape.

> **KEY POINT**
>
> They have a slot or a groove, called the **active site**, into which the **substrate** fits.

The substrate is the substance that is going to react. The reaction then takes place and the **products** leave the enzyme. This explanation for how enzymes work is called the **Lock and Key theory**.

The substrate fits into the active site like a key fitting into a lock.

> **KEY POINT**
>
> Enzymes work best at a particular temperature and pH. This is called the **optimum** temperature or pH.

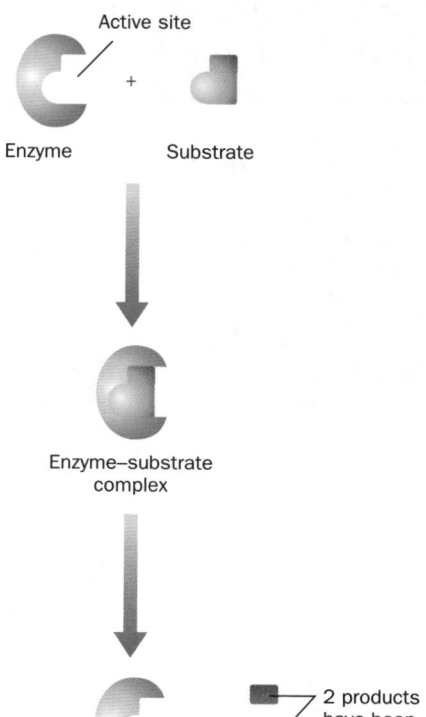

The Lock and Key theory of enzymes.

Active site

Enzyme + Substrate

Enzyme–substrate complex

Ready to be used again + 2 products have been produced

Enzymes have different optimum pH values that depend on where they usually work.

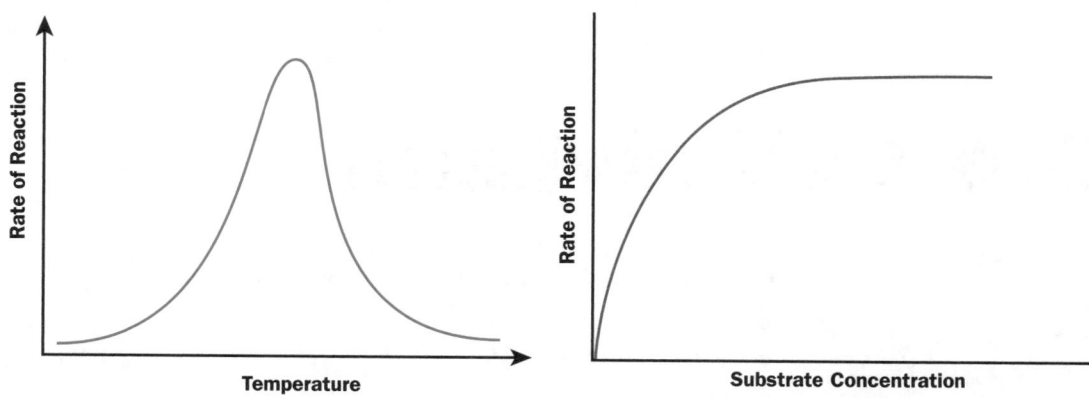

How temperature and substrate concentration affect reaction rate.

Rate of Reaction / Temperature

Rate of Reaction / Substrate Concentration

If the concentration of the substrate is increased then the reaction will be faster up to a certain concentration and then it will level off until all the enzymes are working at their maximum rate. At this point increasing the substrate concentration does not increase the rate of reaction.

Enzyme properties

AQA	B2	✓
OCR A	B4	✓
OCR B	B3	✓
EDEXCEL	B2	✓
WJEC	B2	✓
CCEA	B1	✓

Many candidates lose marks by saying that heat kills enzymes. Remember that enzymes are protein molecules and not living organisms. Say that they are denatured or destroyed, but not killed!

The Lock and Key theory can be used to explain many of the properties of enzymes:

- It explains why an enzyme will only work on one type of substrate. They are described as **specific**. The substrate has to be the right shape to fit into the active site.
- If the temperature is too low, then the substrate and the enzyme molecules will not collide so often and the reaction will slow down. If the shape of the enzyme molecule changes, then the substrate will not easily fit into the active site. This means that the reaction will slow down. High temperatures and extremes of pH may cause this to happen.

KEY POINT

If the shape of the enzyme molecule is irreversibly changed then it is described as being **denatured**.

PROGRESS CHECK

1. Why does a lack of protein stunt growth?
2. Why are enzymes necessary in living organisms?
3. What is the lock and what is the key in the Lock and Key theory?
4. What does the phrase 'optimum temperature' mean?
5. Lipase digests fats, but it will not digest proteins. Explain why this is.
6. Adding vinegar to food can stop the food being digested and spoilt by bacteria and fungi. Explain why this is.

1. Proteins are needed to make key structures inside the body, e.g. bone and so without protein growth will be limited.
2. To allow reactions to be fast enough at body temperature.
3. The lock is the enzyme's active site and the key is the substrate.
4. The temperature at which the reaction occurs at the fastest rate.
5. Lipase has a particular shaped active site that fats will fit into, but not proteins.
6. Vinegar is acidic and so the pH would be too low so the enzymes of the decay organisms would not work.

5.4 Cell division

LEARNING SUMMARY

After studying this section you should be able to:

- describe how DNA is copied
- explain the main differences between mitosis and meiosis
- describe the main sources of genetic variation.

Copying DNA

AQA	B2	✓
OCR A	B5	✓
OCR B	B3	✓
WJEC	B2	✓

Before a cell divides two things must happen. Firstly, new cell organelles such as mitochondria must be made. Secondly, the DNA must copy itself. Watson and Crick realised that the structure of DNA allows this to happen in a rather neat way:

- The double helix of DNA unwinds and the two strands come apart or 'unzip'.
- The bases on each strand attract their complementary bases and so two new molecules are made.

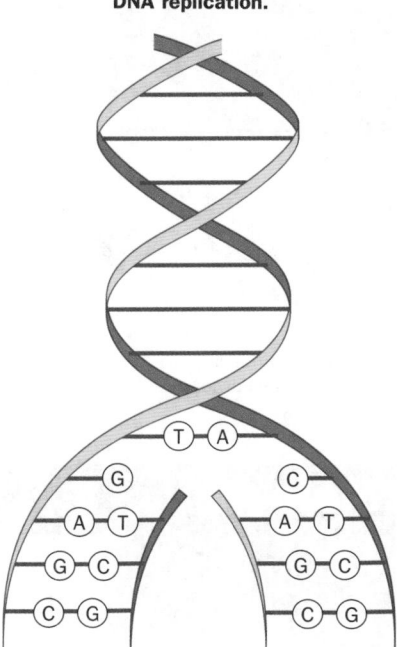

DNA replication.

Types of cell division

AQA	B2	✓
OCR A	B5	✓
OCR B	B3	✓
EDEXCEL	B2	✓
WJEC	B2	✓
CCEA	B2	✓

Cells divide for a number of reasons. There are two types of cell division – **meiosis** and **mitosis** – and they are used for different reasons.

In most questions you will not lose marks if your spelling is a little inaccurate. However, make sure that you spell mitosis and meiosis correctly or the examiner may not know which one you mean.

Edexcel and CCEA candidates need to be able to describe how the sex cells produced by meiosis are adapted for their functions.

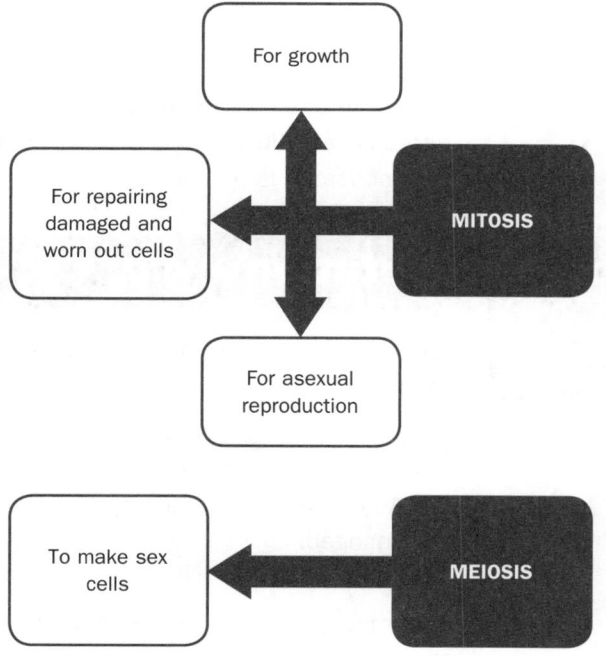

Mitosis

AQA	B2	✓
OCR A	B5	✓
OCR B	B3	✓
EDEXCEL	B2	✓
WJEC	B2	✓
CCEA	B2	✓

In mitosis, two cells are produced from one. As long as the chromosomes have been copied exactly then each new cell will have the same number of chromosomes and therefore the same genetic information as each other and the parent cell.

Mitosis.

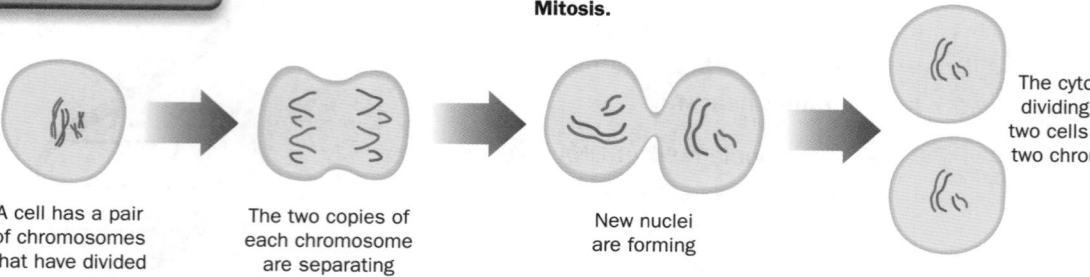

A cell has a pair of chromosomes that have divided

The two copies of each chromosome are separating

New nuclei are forming

The cytoplasm is dividing to make two cells each with two chromosomes

Meiosis

AQA	B2	✓
OCR A	B5	✓
OCR B	B3	✓
EDEXCEL	B2	✓
WJEC	B2	✓
CCEA	B2	✓

In meiosis, the chromosomes are also copied once, but the cell divides twice. This makes four cells each with half the number of chromosomes, one from each pair.

Meiosis.

A cell has a pair of chromosomes each of which has divided

The two chromosomes are separating

Each double stranded chromosome is now split up

Four new cells are formed each with one chromosome

Cells with one chromosome from each pair are called **haploid** and can be used as **gametes**. When two gametes join at fertilisation, the **diploid** or full number of chromosomes is produced.

Variation and mutation

OCR B	B3	✓
EDEXCEL	B2	✓
WJEC	B1	✓

When DNA is copied, before mitosis and meiosis occur, mistakes are sometimes made.

> **KEY POINT**
>
> A **gene mutation** occurs when one of the chemical bases in DNA is changed.

This may mean that a different amino acid is coded for and this can change the protein that is made. When this happens, it is most unlikely to benefit the organism. Either the protein will not be made at all or most likely it will not work properly. Very occasionally, a mutation may be useful and without mutations we would not be here. Mutations occur randomly at a very low rate, but some factors can make them happen more often.

These include:

- ultraviolet in sun light
- X-rays
- chemical mutagens as found in, for example, cigarettes.

Only mutations can produce new genes, but meiosis can recombine them in different orders. Also as a baby can receive any one of the chromosomes in each pair from the mother and any one from the father, the number of possible gene combinations is enormous. This new mixture of genetic information produces a great deal of variation in the offspring. This is why meiosis and sexual reproduction produces so much more variation than asexual reproduction.

> Remember that mitosis can produce cells that are genetically different, but this only happens if there is a mutation. Otherwise, they are genetically identical. Meiosis always produces genetic variation.

PROGRESS CHECK

1. Does a new molecule of DNA have none, one or two original strands?
2. Where in the human body does meiosis occur?
3. The haploid number of chromosomes in humans is 23. What is the diploid number?
4. Write down two differences between mitosis and meiosis.
5. Why is a gene mutation often harmful?
6. Why is it important to make sure that your sunglasses filter out UV light?

6. UV light can be absorbed by DNA and cause mutations.
5. It usually produces a new protein that does not work as well.
4. Meiosis introduces more variation; meiosis makes four cells but mitosis makes two; meiosis makes cells with half the number of chromosomes but mitosis produces cells that have the same number as the parent cells.
3. 46
2. In the ovaries and testes.
1. One.

5.5 Growth and development

LEARNING SUMMARY

After studying this section you should be able to:

- recall the meaning of the term differentiation
- explain the function of stem cells
- describe the main parts of a human growth curve
- describe how different parts of the body grow at different rates
- describe the differences between plant growth and animal growth
- describe ways of measuring growth.

Division and differentiation

AQA	B2	✓
OCR A	B5	✓
OCR B	B3	✓
EDEXCEL	B2	✓
WJEC	B2	✓
CCEA	B1	✓

KEY POINT

When gametes join at fertilisation this produces a single cell called a **zygote**.

The zygote soon starts to divide many times by mitosis to produce many identical cells. These cells then start to become specialised for different jobs.

> **KEY POINT**
>
> The production of different types of cells for different jobs is called **differentiation**.

These differentiated cells can then form tissues and organs.

Stem cells

AQA	B2	✓
OCR A	B5	✓
OCR B	B3	✓
EDEXCEL	B2	✓
WJEC	B2	✓
CCEA	B1	✓

Some cells in the embryo and in the adult keep the ability to form other types of cells. They are called **stem cells**. Scientists are now trying to use stem cells to replace cells that have stopped working or been damaged. This may have the potential to cure a number of conditions.

Uses of stem cells

AQA	B2	✓
OCR A	B5, B7	✓
OCR B	B3	✓
EDEXCEL	B2	✓
WJEC	B2	✓
CCEA	B1	✓

Once a cell has differentiated it does not form other types of cells. Although it has the same genes as all the other cells, many are turned off so it only makes the proteins it needs. Scientists have found a way to switch genes back on and so have been able to clone animals from body cells. This is covered on page 63.

This means that it is now possible to produce embryos that are clones of an animal and to use them to supply embryonic stem cells. There are many different views about the possibility of cloning humans to obtain stem cells.

Human growth curves

OCR B	B3	✓
EDEXCEL	B2	✓
CCEA	B1	✓

Humans grow at different rates at different parts of their lives. This is shown in the graph.

The graph shows that there are two phases of rapid growth, one just after birth and the other in adolescence.

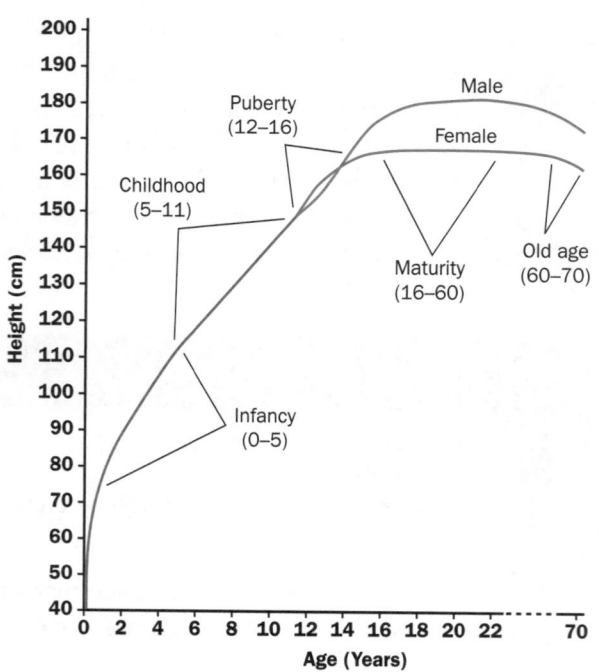

Human growth at different ages.

Growth of different parts of the body

OCR B   B3   ✓

The various parts of the body also grow at different rates at different times.

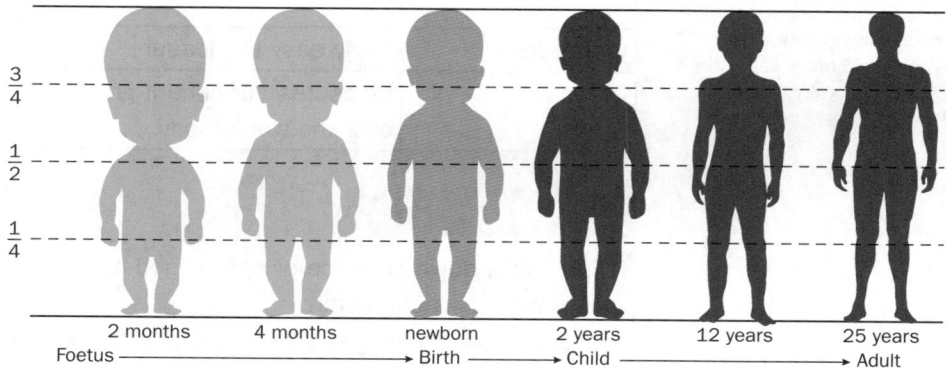

The diagram shows that the head and brain of an early foetus grow very quickly compared to the rest of the body. Later, the body and legs start to grow faster and the brain and head growth slows down into puberty and adulthood.

Plant growth

OCR A	B5	✓
OCR B	B3	✓
EDEXCEL	B2	✓
WJEC	B3	✓
CCEA	B1	✓

Like animals, plants grow by making new cells by mitosis. The cells then differentiate into tissues like xylem and phloem. These tissues then form organs such as roots, leaves and flowers.

Growth in plants is different to animal growth in a number of ways:

- Plant cells enlarge much more than animal cells after they are produced. This increases the size of the plant.
- Cells tend to divide at the ends of roots and shoots. This means that plants grow from their tips.
- Animals usually stop growing when they reach a certain size, but plants carry on growing.
- Many plant cells keep the ability to produce new types of cells, but in animals only stem cells can do this. Plant cells that can produce new types of cells are called **meristematic**.

Measuring growth

OCR B	B3	✓
EDEXCEL	B2	✓
CCEA	B2	✓

Growth can be measured as an increase in **height**, **wet mass** or **dry mass**.

> **KEY POINT**
>
> Dry mass is the best measure of growth.

There are advantages and disadvantages of measuring growth by each method.

> It is easy to use the word weight when talking about measuring growth, but you should really say wet mass or dry mass.

Measurement	Advantage	Disadvantage
Length or height	Easy to measure	Only measures growth in one direction
Wet mass	Fairly easy to measure	Water content can vary
Dry mass	Measures permanent growth over the whole body	Involves removing all the water from an organism

PROGRESS CHECK

1. Write down one type of specialised cell.
2. What is a stem cell?
3. Look at the human growth curve. What does it show about growth in old age?
4. Which parts of a plant contain the main growth areas?
5. Suggest why the head is much larger than the rest of the body when the baby is young.
6. Using dry mass to measure the growth of an organism presents a number of difficulties. Suggest what these difficulties are.

1. Example: Muscle cell.
2. A cell that has not yet differentiated and so can divide to produce any type of cell.
3. Growth becomes negative, i.e. more cells are dying than are being produced.
4. The meristems at the tips of the roots and shoots.
5. The brain needs to develop first to control the other parts of the body.
6. This is destructive and it is difficult to tell when it has all been removed. The organism has to be killed and all water removed in order to measure dry mass. Therefore it is hard to get an idea of growth over time.

5.6 Transport in cells

LEARNING SUMMARY	After studying this section you should be able to:
	• describe the processes of diffusion, osmosis and active transport
	• explain the importance of osmosis in supporting plants
	• describe how to demonstrate osmosis in plant tissue.

Diffusion

AQA	B2	✓
OCR A	B4	✓
OCR B	B4	✓
EDEXCEL	B2	✓
WJEC	B2	✓
CCEA	B2	✓

KEY POINT

Diffusion is the movement of a substance from an area of high concentration to an area of low concentration.

> You must remember that diffusion is high to low concentration. Perhaps the letters DHL might help you to remember this.

Diffusion works because particles are always moving about in a random way. The rate of diffusion can be increased in a number of ways:

Factors that increase diffusion rate.

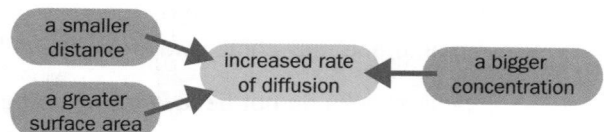

Osmosis

AQA	B3	✓
OCR A	B4	✓
OCR B	B4	✓
EDEXCEL	B2	✓
WJEC	B2	✓

Osmosis is really a special type of diffusion. It involves the diffusion of water.

> **KEY POINT**
>
> Osmosis is the movement of water across a partially permeable membrane from an area of high water concentration to an area of low water concentration.

The cell membrane is an example of a partially permeable membrane. It lets small molecules through, such as water, but stops larger molecules, such as glucose.

Osmosis.

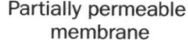
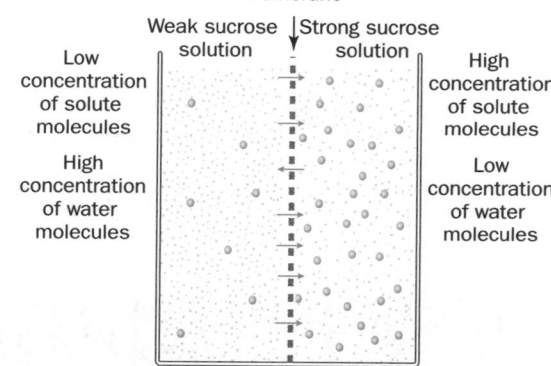
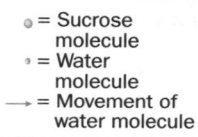

Osmosis and support in plants

OCR B	B4	✓
WJEC	B3	✓
CCEA	B2	✓

When plant cells gain water by osmosis, they swell. The cell wall stops them from bursting. This makes the cell stiff or **turgid**. If a plant cell loses water it goes limp or **flaccid**.

Osmosis in plant cells.

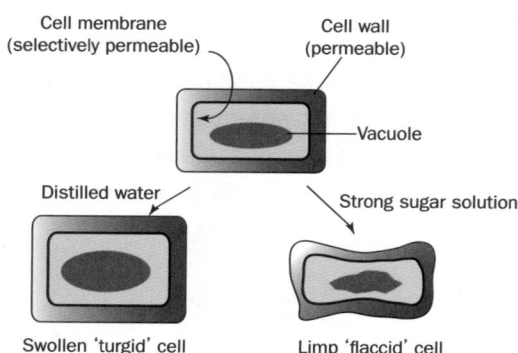

Turgid cells are very important in helping to support plants. Plants with flaccid cells often wilt. Sometimes the cells may lose so much water that the cell membrane may come away from the cell wall. This is called **plasmolysis**.

Animal cells do not behave in the same way because they do not have a cell wall. They will either swell up and burst if they gain water, or shrink if they lose water.

Demonstrating osmosis

AQA	B3	✓
OCR A	B4	✓
OCR B	B4	✓
EDEXCEL	B2	✓
WJEC	B2	✓

It is possible to show how osmosis has occurred by cutting cylinders out of potato and putting them into sugar solutions of different concentrations. If the mass of the chips is taken before and after they are put in the solutions a graph like this can be plotted:

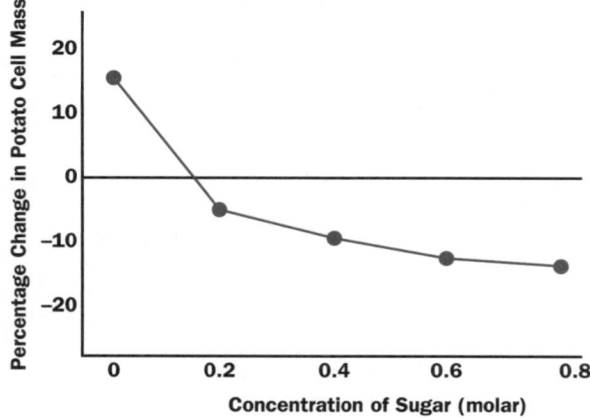

Make sure you can explain the shape of this graph. It might be drawn as the length of the potato chip rather than the mass. It's just the same.

In less concentrated solutions the potato gains water and increases in mass. At high concentrations of sugar the potato loses water and decreases in mass.

Active transport

AQA	B3	✓
OCR A	B4	✓
OCR B	B4	✓
EDEXCEL	B2	✓
WJEC	B2	✓
CCEA	B1	✓

Sometimes substances have to be moved from a place where they are in low concentration to where they are in high concentration. This is in the opposite direction to diffusion and is called **active transport**.

> **KEY POINT**
>
> Active transport is therefore the movement of a substance against the diffusion gradient with the use of energy from respiration.

You can use the phrase 'up or against the diffusion gradient' because this means in the opposite direction to diffusion. Do not say that active transport is 'along' or 'down the diffusion gradient' because this is the wrong way.

Anything that stops respiration will therefore stop active transport. For example, plants take up minerals by active transport. Farmers try and make sure that their soil is not waterlogged because this reduces the oxygen content of the soil, so less oxygen is available to the root cells for respiration. This would therefore reduce the uptake of minerals.

5.7 Respiration

LEARNING SUMMARY

After studying this section you should be able to:

- describe the process of aerobic respiration and ATP production
- explain how anaerobic respiration differs from aerobic respiration
- describe how the rate of respiration can be measured.

What is energy needed for?

AQA	B2	✓
OCR A	B4	✓
OCR B	B3	✓
EDEXCEL	B2	✓
WJEC	B2	✓
CCEA	B1	✓

The energy that is released by respiration can be used for many processes:

- To make large molecules from smaller ones, for example proteins from amino acids.
- To contract the muscles.
- For mammals and birds to keep a constant temperature.
- For active transport.

KEY POINT

Aerobic respiration is when glucose reacts with oxygen to release energy.

> When you are learning the equation for respiration, look at the equation for photosynthesis in the next topic (page 126). Remember that one is just the reverse of the other. Do not try and learn them separately.

Carbon dioxide (CO_2) and water (H_2O) are released as waste products:

glucose + oxygen → carbon dioxide + water + energy

$$C_6H_{12}O_6 + 6O_2 \rightarrow 6CO_2 + 6H_2O \text{ (+ energy released)}$$

The reactions of aerobic respiration take place in mitochondria. All the reactions that occur in our body are called our **metabolism** and so anything that increases our **metabolic rate** will increase our respiration. During exercise the body needs more energy and so the rate of respiration increases.

The breathing rate increases to obtain extra oxygen cend remove carbon dioxide from the body. The heart beats faster so that the blood can transport the oxygen, glucose and carbon dioxide faster. This is why our pulse rate increases.

ATP

OCR B	B3	✓

The energy released by respiration is needed for different processes in different parts of the cell. To make sure that the energy is not lost as heat it is trapped in the bonds of a molecule called ATP. ATP can then pass the energy on to wherever it is needed.

Anaerobic respiration

AQA	B2	✓
OCR A	B4	✓
OCR B	B3, B6	✓
EDEXCEL	B2	✓
WJEC	B2	✓
CCEA	B1	✓

KEY POINT

When not enough oxygen is available, glucose can be broken down by **anaerobic respiration**.

This may happen in muscle cells during hard exercise.

In humans: **glucose → lactic acid + (energy released)**

Being able to respire without oxygen sounds a great idea. However, there are two problems:

- Anaerobic respiration releases much less energy than is released by aerobic respiration.
- Anaerobic respiration produces lactic acid which causes muscle fatigue and pain.

In plants and fungi, such as yeast, anaerobic respiration is often called **fermentation**. It produces different products.

In plants and fungi: **glucose → ethanol + carbon dioxide + (energy released)**

KEY POINT

The build-up of lactic acid in the muscles is called the **oxygen debt** because it needs extra oxygen to be taken in after exercise to deal with it.

Another name for this is **excess post-exercise oxygen consumption (EPOC)**.

The lactic acid is transported to the liver and the heart continues to beat faster to supply the liver with the oxygen needed to break down the lactic acid.

Measuring respiration rate

OCR B	B3	✓

It is possible to measure the respiration rate of organisms by measuring:

- the oxygen consumption
- the carbon dioxide production.

Investigating the rate of oxygen consumption.

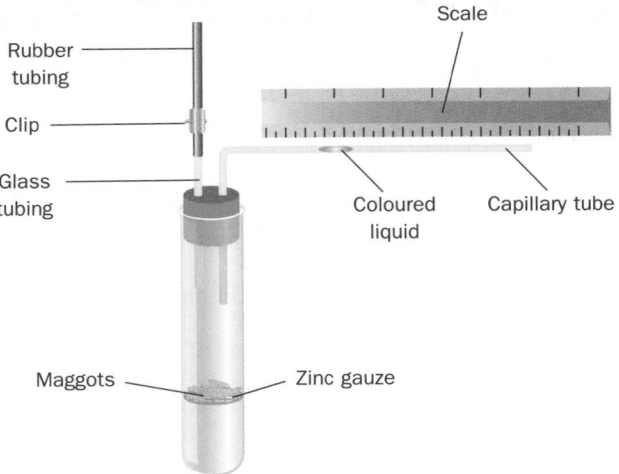

Remember that respiration is controlled by enzymes. This means that any factors that will change the rate of enzyme reactions will also change the rate of respiration.

This apparatus can be used to investigate the rate of oxygen consumption by the maggots. If a liquid that absorbs carbon dioxide is placed in the bottom of the test tube then the coloured liquid will move to the left. Using this apparatus it is possible to investigate the effect of factors such as temperature or pH on the rate of respiration, e.g. by carrying out the experiment at different temperatures. It is also possible to calculate the **respiratory quotient (RQ)** using this formula:

$$RQ = \frac{\text{carbon dioxide produced}}{\text{oxygen used}}$$

The RQ provides useful information about what type of substance is being respired.

PROGRESS CHECK

1. Why do we need to eat more in cold weather?
2. Why do we breathe faster when we exercise?
3. What are the bubbles of gas given off when yeast is fermenting glucose?
4. Why do our muscles hurt when we run a long race?
5. Look at the equation for aerobic respiration using glucose. What would be the RQ of an organism that is respiring only glucose?
6. Wine makers need to control the temperature carefully inside the fermentation tanks when they make wine using fermentation. Explain why.

1. We need to respire more to generate more heat to keep a constant body temperature, so more food is needed to provide glucose for respiration to produce this heat.
2. Respiration rate increases to supply extra energy for muscle contraction. More oxygen is therefore needed and more carbon dioxide needs to be removed.
3. Carbon dioxide.
4. Lactic acid is produced due to anaerobic respiration.
5. 6/6 = 1.0
6. So that the temperature is at an optimum level for the yeast enzymes to produce alcohol.

Sample GCSE questions

1 Grace is doing an experiment to demonstrate osmosis.

She uses a special material called dialysis tubing to make a bag.

The dialysis tubing is selectively permeable.

She half fills the bag with sugar solution and measures its mass.

Grace then lowers the bag into a beaker of sugar solution.

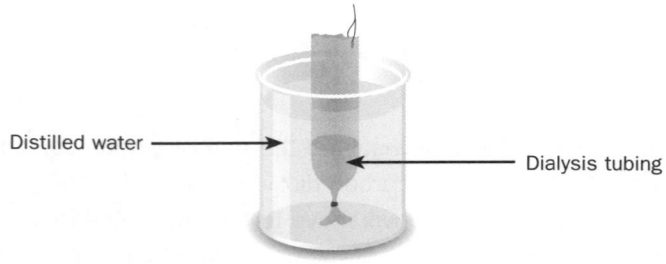

Distilled water ⟶

⟵ Dialysis tubing

(a) The bag is partially permeable. What is meant by partially permeable? **[2]**

This means that some molecules can get through and not others.

> It is important to say that it is usually smaller molecules that can get through.

(b) After thirty minutes Grace takes the bag out of the water and wipes it with a tissue. She then reweighs it.

She repeats this experiment with different concentrations of sugar solution in the beaker.

The solution in the bag is always the same concentration.

Her results are shown in the table.

Concentration of sugar in beaker (mol per dm^3)	Mass of bag before (g)	Mass of bag after (g)	% change in mass
0.0	4.90	5.51	12.40
0.2	4.70	5.10	8.50
0.4	4.80	4.85	1.04
0.6	4.80	4.66	
0.8	5.20	4.81	−7.50

(i) Suggest why Grace wipes the outside of the bag with a tissue before she reweighs it. **[1]**

This is to make sure that she is not weighing any liquid that is on the outside of the bag.

Sample GCSE questions

(ii) Work out the percentage change in mass for the bag in the 0.6 mol per dm³ sugar solution. **[2]**

$$4.66 - 4.80 = \frac{-0.14}{4.80} \times 100 = -2.92\%$$

(iii) Plot the results on the grid.

Finish the graph by drawing the best straight line. **[4]**

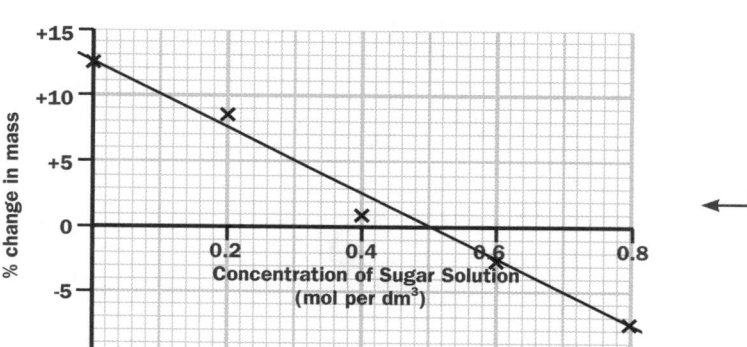

(iv) Describe what happened to the mass of the bag when it was placed into different concentrations of sugar solution. **[2]**

In weak solutions it gained in mass and in strong solutions it lost mass.

(v) Explain why the mass of the bag changed in the different solutions. **[3]**

In weak solutions water passes into the bag by osmosis. This is because the contents are more concentrated.

In stronger solutions water leaves the bag because the gradient is reversed.

(vi) Use the graph to estimate the concentration of sugar solution that was in the bag. Explain how you can tell this from the graph. **[2]**

0.5 mol per dm³

This is the concentration at which no water leaves or enters the bag so the mass does not change.

Make sure that you can do these calculations. Percentage change is the

$$\frac{\text{difference in mass}}{\text{initial mass}} \times 100$$

This has a minus sign because it is decreasing.

When plotting graphs:
- **Make sure you choose a scale that uses at least half the graph paper.**
- **Label the axes with units as well.**

Remember that a best straight line or best curve does not have to go through any of the points.

You should really say what you mean by weak and strong solutions here. Looking at the graph weak is less than 0.5 mol per dm³ and strong is more than 0.5 mol per dm³.

This is correct but it is worth saying that the dialysis tubing is acting as a partially permeable membrane.

This should really say that water is entering and leaving at the same rate.

Exam practice questions

1   An athlete starts to run a race.

(a)   **(i)**   Aerobic respiration is taking place in his muscle cells.

Complete the balanced **symbol** equation for aerobic respiration.

$C_6H_{12}O_6$ + ... $\rightarrow$ .... + $6H_2O$ + (ENERGY RELEASED)   **[2]**

(ii)   The athlete's breathing rate increases during the race.

Explain why.

....

....

....   **[3]**

(b)   Towards the end of the race anaerobic respiration is taking place in the athlete's muscle cells.

Write the **word** equation for anaerobic respiration.

....   **[1]**

(c)   When the athlete has finished the race his breathing rate stays high for some time.

Why does his breathing rate stay high for some time?

....

....

....   **[3]**

Exam practice questions

2 The graphs show how the rates of reaction of two different enzymes alter with temperature.

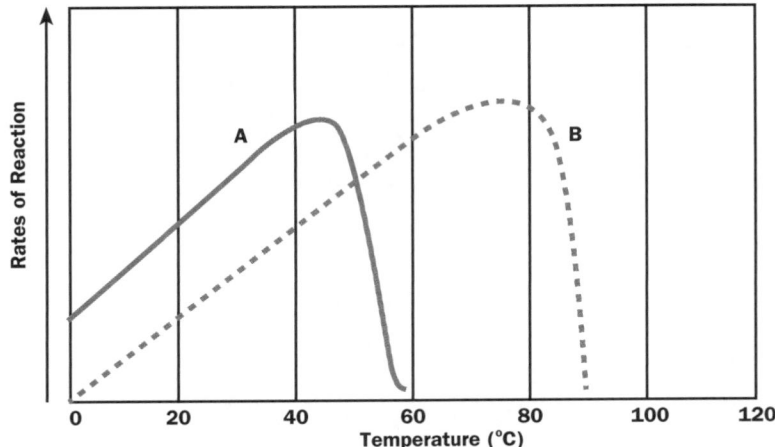

(a) Describe how the rate of reaction with enzyme A alters with temperature.

....

....

.... **[3]**

(b) Explain why the reaction rate alters with temperature.

....

....

.... **[3]**

(c) One of these enzymes is from a human.

The other is from a bacterium that lives in very hot volcanic springs.

Which enzyme is which? Use data from the graph to support your answer.

....

.... **[2]**

Exam practice questions

3 The diagram shows DNA replicating.

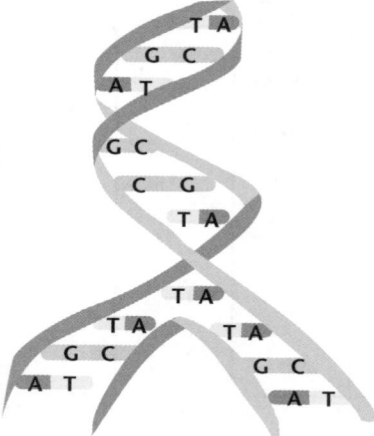

(a) Use the diagram to help you explain how DNA replicates.

....

....

.... **[3]**

(b) DNA replicates before cells divide.

Cells can divide by **mitosis** or by **meiosis**.

Finish this table to show differences between cells made by the two processes.

Feature	Mitosis	Meiosis
Number of cells made from one cell	Two	
Uses of cells that are made		Sex cells
Number of chromosomes made in the cells	Same number as the parent cells	

[3]

(c) The diagram shows a cell dividing by meiosis.

The cell is shown during the first division of meiosis and in the second division.

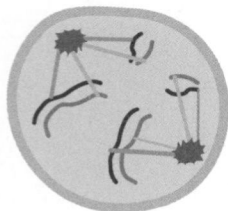 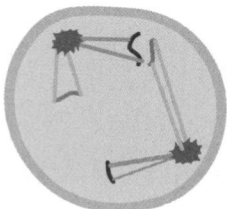

Describe the differences between the two diagrams.

.... **[2]**

6 Sampling organisms

The following topics are covered in this chapter:

- **Sampling techniques**
- **Food production – photosynthesis**
- **Farming techniques**

6.1 Sampling techniques

LEARNING SUMMARY

After studying this section, you should be able to:

- recall the meaning of the terms habitat, population, community and ecosystem
- describe techniques for mapping and sampling an area
- explain what is meant by zonation.

Where do organisms live?

AQA	B2	✓
OCR A	B4	✓
OCR B	B4	✓
EDEXCEL	B2	✓
WJEC	B2	✓
CCEA	B1	✓

Different organisms live in different environments.

- The place where an organism lives is called its **habitat**.
- All the organisms of one type living in a habitat are called a **population**.
- All the populations in a habitat are a **community**.
- An **ecosystem** is all the living and non-living things in a habitat.

Remember that for two organisms to be in the same population they must live in the same habitat and be in the same species. This means that they can successfully mate with each other.

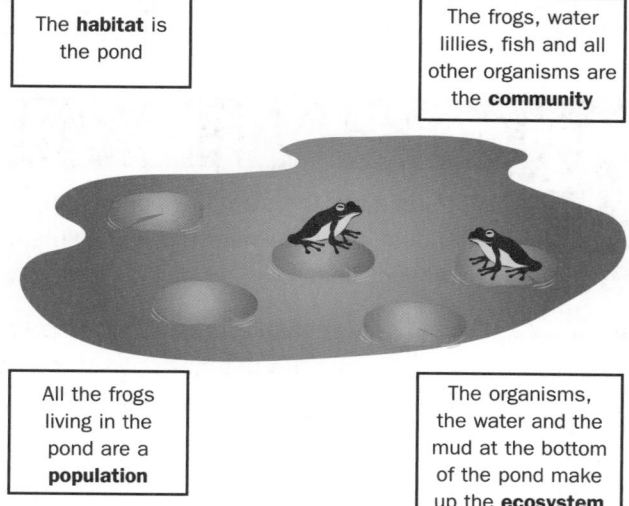

The **habitat** is the pond

The frogs, water lillies, fish and all other organisms are the **community**

All the frogs living in the pond are a **population**

The organisms, the water and the mud at the bottom of the pond make up the **ecosystem**

6 Sampling organisms

Sampling an area

AQA	B2	✓
OCR A	B4	✓
OCR B	B4	✓
EDEXCEL	B2	✓
WJEC	B2	✓
CCEA	B1	✓

It is possible to investigate where organisms live by using various devices.

A **quadrat** is a small square that is put on the ground within which all species of interest are noted or measurements taken. The number of organisms can be counted, or the percentage cover estimated in the quadrat, and the size of the population in the whole area can then be estimated. Often several quadrats are required to determine the estimate.

Quadrats are often used to study plants, but devices such as **pooters**, **nets** and **pitfall traps** can be used to sample animal populations.

It is easy to try and estimate how many of one type of plant live in a habitat:

- Work out the area of the whole habitat.
- Sample a small area using several quadrats and count out how many plants are present.
- Scale up this number to give an estimate of the population in the whole habitat.

Working out the population of animals is harder because they do not keep still to be counted!

We can use a technique called **mark–recapture**:

- The organisms, such as snails, are captured, unharmed.
- They are counted and then marked in some way, for example the snail can be marked with a dot of non-toxic paint.
- They are released.
- Some time later the process of capturing is repeated and another count is made.
- This count includes the number of marked animals and the number unmarked.

To work out the estimate of the population a formula is used.

Population size is:

$$\frac{\text{number in 1st sample} \times \text{number in 2nd sample}}{\text{number in 2nd sample previously marked}}$$

> Remember in all sampling questions, the more samples that you take in an area, the more accurate the estimate of the whole area will be.

Mapping an area

AQA	B2	✓
OCR A	B4	✓
OCR B	B4	✓
EDEXCEL	B2	✓
WJEC	B2	✓
CCEA	B1	✓

To estimate the size of a population in an area we can use quadrats put down at random. To see where the organisms live in a habitat we can use a **transect line**:

- A tape measure is put down in a line across the habitat.
- Quadrats are put down at set intervals along the tape.
- The organisms in the quadrats are then counted.

Zonation

OCR B   B4   ✓

In some habitats a transect line can produce interesting results. Different organisms live at different points along the line. This is because there is a change in environmental conditions along the line. This is called **zonation**.

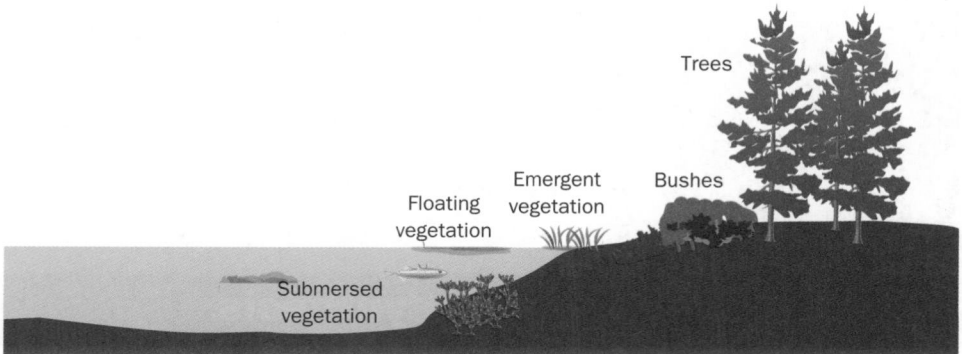

Trees

Emergent vegetation

Bushes

Floating vegetation

Submersed vegetation

An example of zonation is found in a pond. Different plants can grow at different distances into the pond. This is due to the amount of water in the soil.

Artificial ecosystems

OCR B   B4   ✓

Our planet has a range of different ecosystems. Some of these are **natural**, such as woodland and lakes. Others are **artificial** and have been created by man, such as fish farms, greenhouses and fields of crops.

Artificial ecosystems usually have less variety of organisms living there (less biodiversity). This may be caused by the use of chemicals such as weed killers, pesticides and fertilisers.

PROGRESS CHECK

1. What name is given to all the rabbits living in the same field?
2. What device would you use to sample **a)** daisies on a field, **b)** butterflies and **c)** woodlice?
3. Five daisy plants are found in 0.25 m². How many would there be in the whole 100 m² field?
4. Why is it best to sample several areas in the field and take an average?
5. 30 snails are collected in an area, marked and released. When another sample is captured there are 35 snails and 5 are marked. What is an estimate of the snail population?
6. Different animals live on different parts of a rocky shore on the way down to the sea. Use the idea of tides to explain why the animals show zonation.

6. Some animals can survive more time out of the water than others and so can live further up the shore.
5. 210
4. One area of the field may not be representative of all the areas.
3. 20 in 1 m² so 2000 in the field.
2. **a)** quadrat; **b)** net; **c)** pitfall trap.
1. A population.

6.2 Food production – photosynthesis

LEARNING SUMMARY

After studying this section, you should be able to:

- recall the word and symbol equations for photosynthesis
- describe leaf structure
- describe how the understanding of photosynthesis has developed
- explain what is meant by limiting factors.

The reactions of photosynthesis

AQA	B2	✓
OCR A	B4	✓
OCR B	B4	✓
EDEXCEL	B2	✓
WJEC	B2	✓
CCEA	B1	✓

Try this for remembering the equation for photosynthesis:

Certain **W**orms **E**at **G**rass **O**utside (**c**arbon **d**ioxide **w**ater **e**nergy **g**lucose **o**xygen).

> **KEY POINT**
>
> Plants make their own food by a process called **photosynthesis**.

They take in carbon dioxide and water and turn it into sugar, releasing oxygen as a waste product. The process needs the energy from sunlight and this is trapped by the green pigment **chlorophyll**.

(light energy)

carbon dioxide + water → glucose + oxygen

(chlorophyll)

$$6CO_2 + 6H_2O \rightarrow C_6H_{12}O_6 + 6O_2$$

Where does photosynthesis happen?

AQA	B2	✓
OCR A	B4	✓
OCR B	B4	✓
EDEXCEL	B2	✓
CCEA	B1	✓

Photosynthesis occurs mainly in the leaves.

Cross section of a leaf.

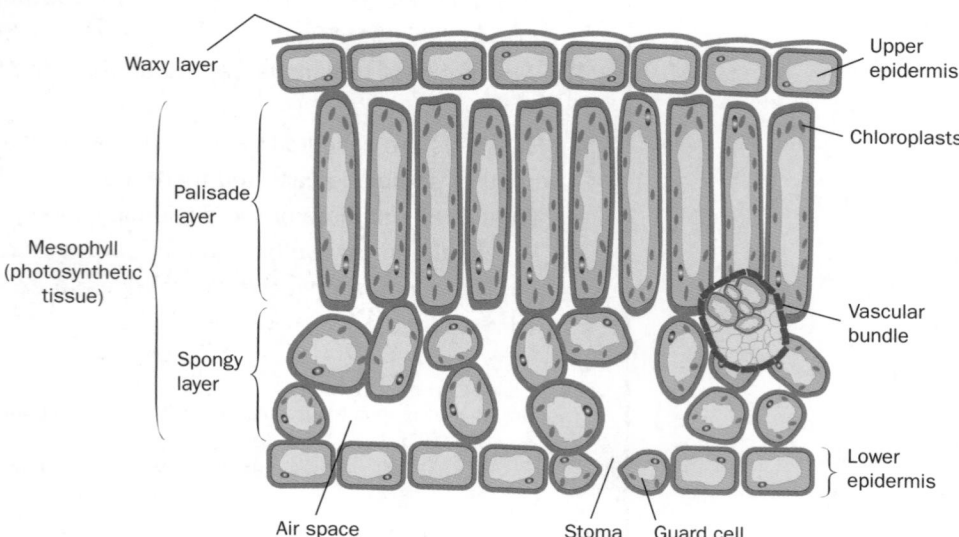

Waxy layer · Upper epidermis · Chloroplasts · Palisade layer · Mesophyll (photosynthetic tissue) · Spongy layer · Vascular bundle · Air space · Stoma · Guard cell · Lower epidermis

The leaves are specially adapted for photosynthesis in a number of ways:

Adaptation	How it helps photosynthesis
A broad shape	Provides a large surface area to absorb light and CO_2
A flat shape	The gases do not have too far to diffuse
Contains a network of veins	Supplies water from the roots and takes away the products
Contains many chloroplasts in the palisade layer near the top	This traps the maximum amount of light
Pores called stomata and air spaces	They allow gases to diffuse into the leaf and reach the cells

Photosynthesis experiments

AQA	B2	✓
OCR A	B4	✓
OCR B	B4	✓
EDEXCEL	B2	✓
WJEC	B2	✓
CCEA	B1	✓

'How Science Works' questions may ask for examples of how ideas in science have changed over the years. The discoveries made in photosynthesis are good examples to use.

The understanding of the process of photosynthesis has changed considerably during history:

- Greek scientists thought that plants gained mass only by taking in minerals from the soil.
- Van Helmont in the seventeenth century worked out that plant growth cannot be solely due to minerals from the soil. He found that the mass gained by a plant was more than the mass lost by the soil.
- In the eighteenth century Priestley showed that oxygen is produced by plants.

More modern experiments using **isotopes** have increased our understanding of photosynthesis. Isotopes of carbon can be used that behave in the same way chemically, but can be followed in the reactions because they are radioactive.

These experiments have shown that photosynthesis is a two stage process:

- Light energy is used to split water, releasing oxygen gas and hydrogen atoms.
- Carbon dioxide gas combines with the hydrogen to make glucose and water.

The rate of photosynthesis can be increased by providing:

- more light
- more carbon dioxide
- an optimum temperature.

Any of these factors can be **limiting factors**.

> **KEY POINT**
>
> A limiting factor is something that controls how fast a reaction will occur.

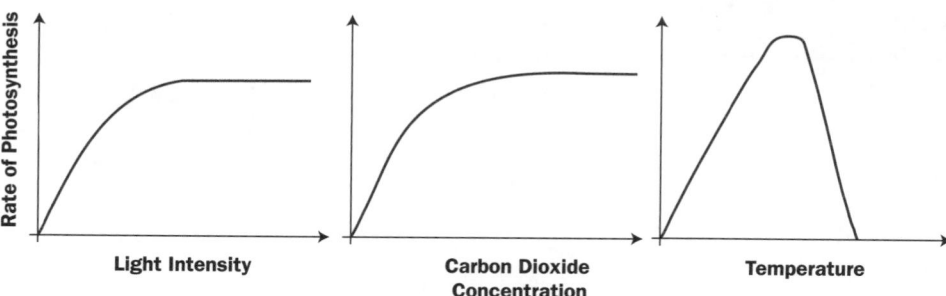

If more light is provided, it will increase photosynthesis because more energy is available.

After a certain point something else will limit the rate.

More carbon dioxide will again increase the rate up to a point because more raw materials are present.

Increasing the temperature will make enzymes work faster, but high temperatures denature enzymes.

PROGRESS CHECK

1. What is the job of chlorophyll in photosynthesis?
2. Which cells in the leaf have most chloroplasts?
3. What are stomata?
4. Why do leaves have veins?
5. In the reactions of photosynthesis is oxygen released from water, carbon dioxide or both?
6. Why do high temperatures stop photosynthesis happening?

6. High temperatures will denature the enzymes that control the reactions of photosynthesis.
5. It is released from water.
4. To supply water for photosynthesis and to take away the sugars that are produced from photosynthesis.
3. Structures containing leaf pores that allow gases in and out of the leaf.
2. Palisade mesophyll.
1. To absorb the energy of sunlight.

6.3 Farming techniques

LEARNING SUMMARY

After studying this section, you should be able to:
- describe the uses that plants have for glucose
- describe methods for intensive food production
- explain how intensive food production increases yields
- describe how organic farming differs from intensive farming
- describe the main methods for food preservation.

Plants need minerals

AQA	B2	✓
OCR A	B4	✓
OCR B	B4	✓
WJEC	B2	✓
CCEA	B1	✓

Once plants have made sugars, such as glucose by photosynthesis, they can convert the glucose into many different things that they need to grow:

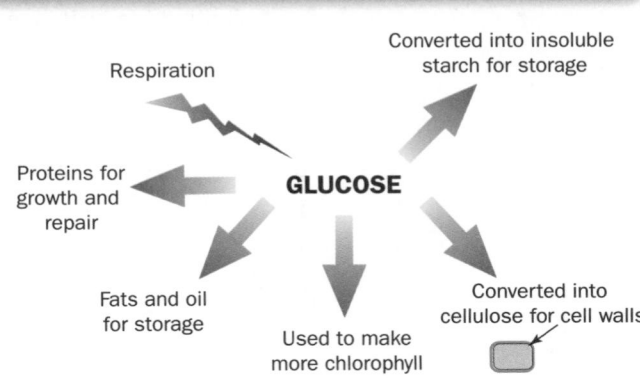

Fertilisers are labelled with figures called an NPK value. This gives the ratio of nitrates, phosphates and potassium in the fertiliser.

To produce these chemicals, plants need various minerals from the soil:

- **Nitrates** as a supply of nitrogen to make amino acids and proteins.
- **Phosphates** to supply phosphorus to make DNA and cell membranes.
- **Potassium** to help enzymes in respiration and photosynthesis.
- **Magnesium** to make chlorophyll.

Without these minerals plants do not grow properly. So farmers must make sure that they are available in the soil.

Intensive food production

OCR B	B4	✓
WJEC	B1	✓
CCEA	B1	✓

The human population is increasing and so there is a greater demand for food. This means that many farmers now use **intensive farming** methods.

> **KEY POINT**
>
> Intensive farming means trying to obtain as much food as possible from the land.

There are a number of different food production systems that use intensive methods:

Fish farming

Fish are kept in enclosures away from predators. Their food supply and pests are controlled.

Glasshouses

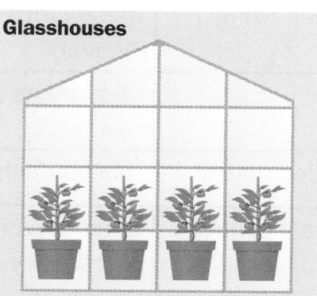

Plants can be grown in areas where the climate would not be suitable. They can also produce crops at different times of the year.

Hydroponics

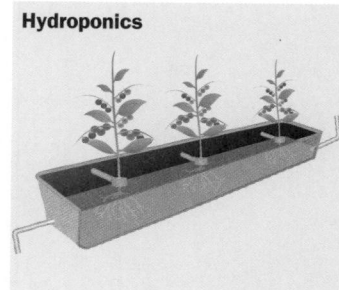

Plants are grown without soil. They need extra support but their mineral supply and pests are controlled.

Energy flow in food production

| AQA | B3 | ✓ |
| OCR B | B4 | ✓ |

Farmers use a number of intensive farming techniques to help increase their yield:

'How Science Works' questions may ask for the arguments for and against methods of intensive farming. Make sure that you can give both sides of the argument.

Using pesticides to kill pests that might eat the crop

Using herbicides to kill weeds that would compete with the crop

INTENSIVE FARMING

Keep animals indoors so that they do not waste energy keeping warm or moving about

Provide the plants with chemical fertilisers for growth

One argument is that the damage caused by some of these techniques does not justify the increase in food production.

Organic food production

OCR B   B4   ✓
WJEC   B2   ✓

Many people think that intensive farming is harmful to the environment and cruel to animals.

> **KEY POINT**
>
> Farming that does not use the intensive methods is called **organic farming**.

Organic farming uses a number of different techniques:

Technique	Details
Use of manure and compost	These provide minerals for the plant instead of using chemical fertilisers.
Crop rotation	Farmers do not grow the same crop in the same field year after year. This stops the build-up of pests and often reduces nutrient depletion of the soil.
Use of nitrogen fixing crops	These crops contain bacteria that add minerals to the soil.
Weeding	This means that chemical herbicides are not needed.
Varying planting times	This can help to avoid times that pests are active.
Using biological control	Farmers can use living organisms to help to control pests. They may eat them or cause disease.

Preserving food

OCR B   B4   ✓

Although gardeners want decay to happen in their compost heaps, people do not want their food to decay before they can eat it.

> **KEY POINT**
>
> **Food preservation** methods reduce the rate of decay of foods.

There are many ways to preserve food. Most stop decay by taking away one of the factors that decomposers need:

Preservation method	How it is done	How does it work?
Canning	Food is heated in a can and the can is sealed.	The high temperature kills the microorganisms and oxygen cannot get into the can after it is sealed.
Cooling	Food is kept in a refrigerator at about 5°C.	The growth and respiration of the decomposers slow down at low temperature.
Freezing	Food is kept in a freezer at about −18°C.	The decomposers cannot respire or reproduce.
Drying	Dry air is passed over the food.	Microorganisms cannot respire or reproduce without water.
Adding salt or sugar	Food is soaked in a sugar solution or packed in salt.	The sugar or salt draws water out of the decomposers.
Adding vinegar	The food is soaked in vinegar.	The vinegar is too acidic for the decomposers.

Look back at pages 85–86 which cover decay. Make sure that you can explain how a food preservation method stops decay.

Modern methods of food packaging can help to increase the shelf life of food and to detect contaminants. Some of these methods involve the use of **nanotechnology**. For example, nanosensors in plastic packaging can detect gases given off by food when it spoils and the packaging changes colour as an alert that the food has gone bad.

PROGRESS CHECK

1. Why do plants need nitrates?
2. Why does a plant look yellow if grown with a lack of magnesium?
3. What is hydroponics?
4. Why does food still go bad in a refrigerator?
5. Suggest one problem with using large quantities of chemical pesticides to kill insect pests.
6. Why is the food brought to pigs in intensive farming rather than letting them find food?

1. To produce amino acids and therefore proteins.
2. Chlorophyll cannot be made as chlorophyll contains magnesium.
3. Growing plants without the use of soil, just in water.
4. The temperature is not low enough to completely stop the growth of microbes that cause decay.
5. Often they kill the natural predators of the pests too and so increase the problem. They also contaminate the environment and reduce biodiversity.
6. It prevents loss of energy in pig movement therefore leaving more energy for growth.

Sample GCSE questions

1 **(a)** Plants absorb energy from sunlight to make food by photosynthesis.

Complete the balanced symbol equation for photosynthesis. **[2]**

(light energy)

$$6CO_2 + 6H_2O \rightarrow C_6H_{12}O_6 + 6O_2$$

(chlorophyll)

> Make sure that you follow the same rules as for chemistry equations and use the correct size numbers and letters.

(b) The chloroplasts that absorb light energy for photosynthesis are found mainly near the top surface of the leaf.

Stomata that absorb carbon dioxide are found on the lower surface.

Explain these two observations. **[3]**

The chloroplasts are towards the top of the leaf because that is where the light energy is at its greatest.

Stomata, however, are on the underside to try and reduce water loss. It is cooler and there is less air movement underneath the leaf so less water is lost.

> A good answer. You could use the term transpiration to describe the loss of water.

(c) Plants only photosynthesise during the day but respire all the time.

How can plants manage to make more glucose than they use? **[2]**

During the daytime the rate of photosynthesis is usually much greater than the rate of respiration.

At night only respiration occurs but the sugar used up is less than the surplus made during the day.

> Remember that plants respire day and night. Many candidates think that they only respire at night.

(d) The graph shows the rate of photosynthesis as light intensity changes at three different concentrations of carbon dioxide.

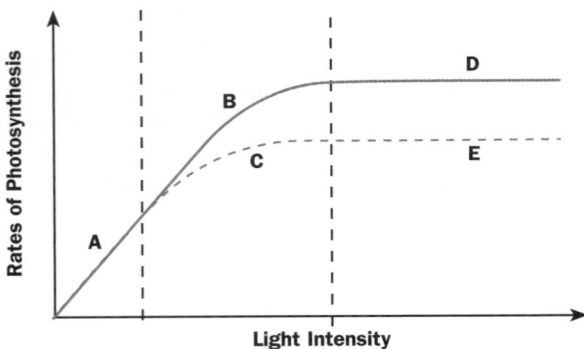

Which letter (**A**, **B**, **C**, **D** or **E**) shows where carbon dioxide is the only limiting factor? Explain your answer. **[3]**

E. It is the only section of the graph where increasing the concentration of carbon dioxide increases the rate but increasing light intensity does not.

> Limiting factors is a difficult idea. Remember that a factor is limiting if increasing that factor will increase the rate of photosynthesis.

Exam practice questions

1   Kane investigates a pond.

He observes various plants and animals that live in the pond.

(a)   Using the area that Kane investigated, explain what is meant by the following terms.

(i)   habitat

....

.... **[1]**

(ii)   population

....

.... **[1]**

(iii)   ecosystem

....

.... **[1]**

(b)   Kane wants to work out how many snails there are in the pond.

He decides to sample one small area of 1 m^2.

In this area he catches five snails.

(i)   What device might he use to catch the snails?

.... **[1]**

(ii)   Kane measures the diameter of the circular pond as 15 metres.

Work out an estimate of the number of snails in the whole pond.

answer = ... snails **[2]**

(iii)   Kane then tries another method.

He samples in five different areas, catches and marks 30 snails.

He releases them and then samples in the same way the next day.

He catches 29 snails and two were marked.

Use this formula to estimate the number of snails in the pond:

$$\text{number of snails} = \frac{\text{number in first sample x number in second sample}}{\text{number in the second sample that have been marked}}$$

answer = ... snails **[2]**

(iv)   Which of Kane's two estimates is likely to be the most accurate?

Explain why.

....

.... **[2]**

Exam practice questions

2 In the seventeenth century a Dutch scientist called van Helmont carried out a famous experiment.

At that time people thought that plants obtained their food from the soil.

He grew a tree in a pot of soil, supplying it only with rain water.

The diagram shows the measurements that he took.

2 kg tree + 100 kg soil   Tree grows for 5 years with rain water   80 kg tree + 99 kg soil

(a) (i) Explain how van Helmont used these results to disprove the idea that plants gain their food entirely from the soil.

....

....

.... **[2]**

(ii) Van Helmont concluded that the tree gained in mass entirely from the rain water.

To what extent is this true?

....

....

.... **[3]**

(b) In 1953 an American scientist called Melvin Calvin worked out many of the reactions of photosynthesis.

How would he communicate his ideas to other scientists and how would these methods differ from those used by van Helmont?

....

....

....

.... **[3]**

Exam practice questions

3 Red spiders are pests of plants that grow in greenhouses such as tomatoes.

(a) The spiders can be killed by spraying with an insecticide.

(i) Some gardeners do not want to spray their tomato plants with insecticide.

Suggest a reason why.

....

.... **[1]**

(ii) Over a number of years an insecticide may become less effective in killing the spiders.

Suggest why this might be.

.... **[1]**

(b) Another way to kill the spiders is to buy some mites.

The mites can be released into the greenhouse to eat the spiders.

(i) The use of the mite is an example of a different type of pest control.

What is it called?

.... **[1]**

(ii) Scientists have to be careful when they introduce this type of control.

Explain why.

....

.... **[2]**

(c) Some farmers grow tomatoes with their roots in liquid rather than in soil.

What name is given to this method of growing crops and what are the advantages?

....

....

....

.... **[3]**

7 Physiology

The following topics are covered in this chapter:

- **Transport in animals**
- **Transport in plants**
- **Digestion and absorption**

7.1 Transport in animals

LEARNING SUMMARY

After studying this section, you should be able to:

- describe the roles of the different cells found in blood
- compare the structure and function of the different types of blood vessels
- describe the structure of the heart and the double circulation.

Blood

AQA	B3	✓
OCR A	B7	✓
OCR B	B3	✓
EDEXCEL	B2	✓
WJEC	B3	✓
CCEA	B2	✓

Blood is made up of a liquid called **plasma**.

> **KEY POINT**
>
> Plasma carries chemicals such as dissolved food, hormones, antibodies and waste products around the body.

Cells are also carried in the plasma. They are adapted for different jobs.

Red blood cells are shaped like a biconcave disc. They contain haemoglobin which can carry oxygen around the body.

A red blood cell.

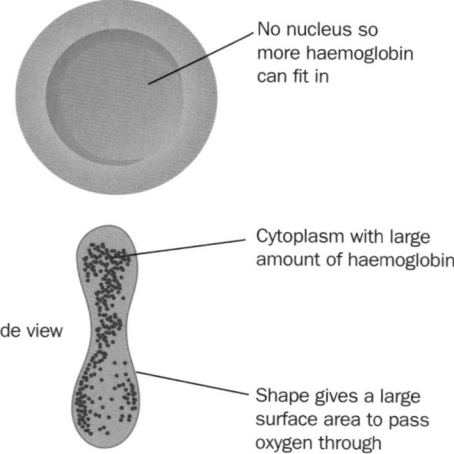

No nucleus so more haemoglobin can fit in

Cytoplasm with large amount of haemoglobin

Side view

Shape gives a large surface area to pass oxygen through

> The shape of red blood cells does not allow them to carry more oxygen, but the increased surface area to volume ratio lets them gain or lose it quicker.

> **KEY POINT**
>
> The haemoglobin in red blood cells reacts with oxygen in the lungs forming **oxyhaemoglobin**.

In the tissues the reverse of this reaction happens and oxygen is released:

haemoglobin + oxygen $\rightleftharpoons$ oxyhaemoglobin

> **KEY POINT**
>
> **White blood cells** can change shape to engulf and destroy disease organisms. They can also produce antibodies.
>
> **Platelets** are responsible for clotting the blood.

The blood is carried around the body in **arteries**, **veins** and **capillaries**.

Arteries	Veins	Capillaries
Carry blood away from the heart	Carry blood back to the heart	Join arteries to veins.
Have thick, muscular walls because the blood is under high pressure	Have valves and a wide lumen because the blood is under low pressure.	Have permeable walls so that substances can pass in and out to the tissues.

> You need to remember Arteries carry blood away from the heart and veINs back INto the heart.

The heart

AQA	B3	✓
OCR A	B2, B7	✓
OCR B	B3, B5	✓
EDEXCEL	B2	✓
WJEC	B3	✓
CCEA	B2	✓

> **KEY POINT**
>
> The heart is made up of four chambers.

The top two chambers are called **atria** and they receive blood from veins. The bottom two chambers are **ventricles**. They pump the blood out into arteries.

The top two chambers, the atria, fill up with blood returning in the **vena cava** and **pulmonary veins**. The two atria then contract together and pump the blood down into the ventricles. The two ventricles then contract pumping blood out into the **aorta** and **pulmonary artery** at high pressure.

> Make sure that you can spot that the muscle wall of the left ventricle is always thicker than the right ventricle. This is because it has to pump blood all round the body compared to the short distance to the lungs.

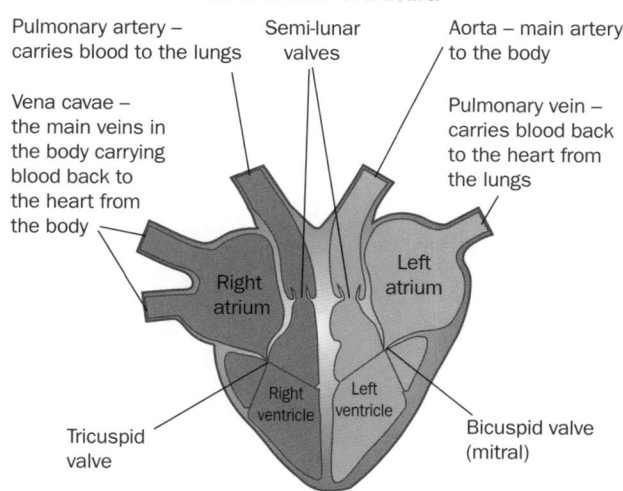

Cross section of a heart.

Pulmonary artery – carries blood to the lungs

Semi-lunar valves

Aorta – main artery to the body

Vena cavae – the main veins in the body carrying blood back to the heart from the body

Pulmonary vein – carries blood back to the heart from the lungs

Right atrium

Left atrium

Right ventricle

Left ventricle

Tricuspid valve

Bicuspid valve (mitral)

> **KEY POINT**
>
> In the heart there are two sets of valves, whose function is to prevent blood flowing backwards.

In between the atria and the ventricles are the **bicuspid** and **tricuspid valves**. These valves stop blood flowing back into the atria when the ventricles contract. The pressure of blood closes the flaps of the valves and the tendons stop the flaps turning inside out. There are also **semi-lunar** valves between the ventricles and the arteries.

A double circulation

OCR B	B3, B5 ✓
WJEC	B3 ✓

> **KEY POINT**
>
> Mammals have a **double circulation**. This means that the blood has to pass through the heart twice on each circuit of the body.

Deoxygenated blood is pumped to the lungs and the oxygenated blood returns to the heart to be pumped to the body.

The advantage of this system is that the pressure of the blood stays quite high and so it can flow faster around the body. Because of the double circulation the heart is really two pumps in one:

- The right side pumps the blood to the lungs.
- The left side pumps it to the rest of the body.

PROGRESS CHECK

1. What is the job of platelets?
2. Why do red blood cells lack a nucleus?
3. Why do veins have valves?
4. What blood vessel carries blood from the heart to the lungs?
5. Why is the right side of the heart coloured blue in the diagram?
6. Some people have a defect in the bicuspid valve. Explain why this can lead to a build up of blood in the blood vessels of the lungs.

1. To clot the blood.
2. To fit more haemoglobin in and so carry more oxygen.
3. To stop the blood flowing backwards as the pressure is low.
4. Pulmonary artery.
5. It contains deoxygenated blood (which is dark red, not blue).
6. When the left ventricle contracts some of the blood goes back into the left atrium rather than out into the aorta. This causes a backlog of blood in the veins coming back from the lungs.

7.2 Transport in plants

After studying this section, you should be able to:

- describe the position and function of xylem and phloem
- explain how water moves through a plant
- explain the factors that affect transpiration rate and how it can be reduced.

Xylem and phloem

AQA	B3	✓
OCR B	B4	✓
EDEXCEL	B2	✓
WJEC	B3	✓
CCEA	B2	✓

KEY POINT

Plants have two different tissues that are used to transport substances. They are called **xylem** and **phloem**.

Xylem vessels and phloem tubes are gathered together into collections called **vascular bundles**. They are found in different regions of the leaf, stem and root.

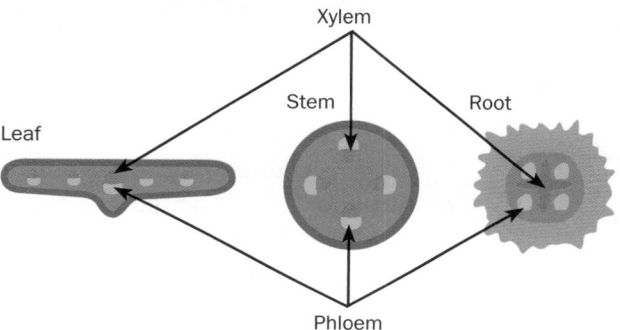

Xylem vessels and phloem tubes are different in structure and do different jobs:

Xylem	Phloem
Carries water and minerals from roots to the leaves	Carries dissolved food substances both up and down the plant
The movement of water up the plant and out of the leaves is called **transpiration**	The movement of the dissolved food is called **translocation**
Made of vessels which are hollow tubes made of thickened dead cells	Made of columns of living cells

Remember phloem for food and xylem for water.

The movement of water

AQA	B3	✓
OCR B	B4	✓
EDEXCEL	B2	✓
WJEC	B3	✓
CCEA	B2	✓

KEY POINT

Water enters the plant through the **root hairs**, by **osmosis**.

The root hair cells increase the surface area for the absorption of water.

> Remember that it is osmosis that brings the water into the leaf and into the xylem, but water does not move up the xylem by osmosis. It is 'sucked up' by evaporation from the leaves.

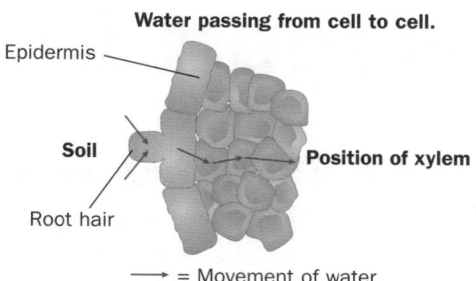

Water passing from cell to cell.

Epidermis

Soil

Root hair

Position of xylem

⟶ = Movement of water

Water then passes from cell to cell by osmosis until it reaches the centre of the root. The water enters xylem vessels in the root and then travels up the stem. Water enters the leaves and evaporates. It then passes through the **stomata** by **diffusion**.

> **KEY POINT**
>
> This loss of water is called **transpiration** and it helps to pull water up the xylem vessels.

Various environmental conditions can affect the transpiration rate.

Transpiration rate

AQA	B3	✓
OCR B	B4	✓
WJEC	B3	✓
CCEA	B2	✓

The rate of transpiration depends on a number of factors:

- **Temperature** – warm weather increases the kinetic energy of the water molecules so they move out of the leaf faster.
- **Humidity** – damp air reduces the concentration gradient so the water molecules leave the leaf more slowly.
- **Wind** – the wind blows away the water molecules so that a large diffusion gradient is maintained.
- **Light** – light causes the stomata to open and so more water is lost.

> If a question asks *Give one factor that increases transpiration rate*, make sure that you write 'an increase in temperature' not just 'temperature'. Many candidates lose marks in this way.

Adaptations to reduce water loss

| AQA | B3 | ✓ |
| OCR B | B4 | ✓ |

When plants are short of water, they do not want to waste it through transpiration. The trouble is they need to let carbon dioxide in, so water will always be able to get out. Water loss is kept as low as possible in several ways:

- Photosynthesis only occurs during the day, so the stomata close at night to reduce water loss. The guard cells lose water by osmosis and become flaccid. This closes the stoma.
- The stomata are on the underside of the leaf. This reduces water loss because they are away from direct sunlight and protected from the wind.
- The top surface of the leaf, facing the Sun, is often covered with a protective waxy layer.

Although transpiration is kept as low as possible, it does help plants by cooling them down and supplying leaves with minerals. It also provides water for support and photosynthesis.

7.3 Digestion and absorption

After studying this section, you should be able to:

LEARNING SUMMARY	• describe the position and function of the main parts of the digestive system • describe how the products of digestion are absorbed • describe some other uses of digestive enzymes.

Digestion

AQA	B2	✓
OCR B	B5	✓
EDEXCEL	B2	✓
WJEC	B2	✓
CCEA	B1	✓

KEY POINT

The job of the digestive system is to break down large food molecules into small soluble molecules. This is called digestion.

Digestion happens in two main ways – physical and chemical digestion.

Physical digestion occurs in the mouth where the teeth break up the food into smaller pieces. Chemical digestion is caused by digestive enzymes that are released at various points along the digestive system. Most enzymes work inside cells controlling reactions. Some enzymes pass out of cells and work in the digestive system. These enzymes digest our food, making the molecules small enough to be absorbed.

PROGRESS CHECK

1. In which direction in a plant stem does water and minerals move?
2. What is translocation?
3. Where in a plant root is xylem found?
4. What is the function of root hair cells?
5. Why is it impossible for plants to prevent all water loss from the leaves?
6. What causes stomata to close when a plant wilts?

1. Up, towards the leaves.
2. The movement of dissolved food through the phloem.
3. In the centre, in a star shape.
4. To increase the surface area for water absorption.
5. Because carbon dioxide must be allowed in for photosynthesis.
6. The guard cells lose water by osmosis and so the cells become flaccid, straightening up and closing the pore.

The digestive system.

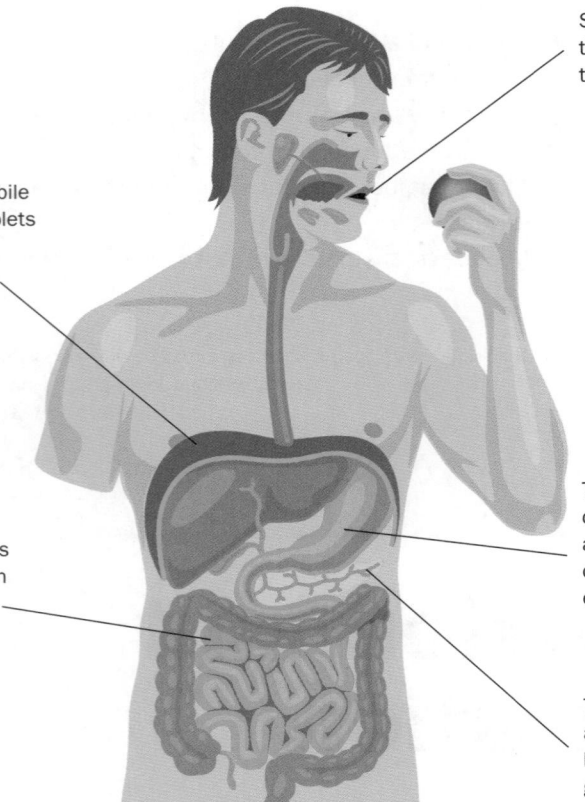

Saliva is released into the mouth from the salivary glands. It contains amylase to break down starch to maltose.

The liver makes bile that contains bile salts. They break the large fat droplets down into smaller droplets. Bile is stored in the gall bladder.

The stomach makes gastric juice, containing protease and hydrochloric acid. The acid kills microbes and creates the best pH for the protease to digest proteins.

The small intestine makes enzymes such as maltase. This breaks down maltose to glucose.

The pancreas makes more protease and amylase. It also makes lipase to break down the fats to fatty acids and glycerol.

The food is moved along the gut by contractions of the muscle in the lining of the intestine. This process is called **peristalsis**.

The rate of digestion

OCR B	B5	✓
EDEXCEL	B2	✓
WJEC	B2	✓
CCEA	B1	✓

To make the digestive enzymes work at an optimum rate the digestive system provides the best conditions:

- Each enzyme has a different optimum pH. Protease in the stomach works best at about pH 2, but a different protease made by the pancreas works best at about pH 9.
- Physical digestion helps to break the food into smaller particles, thereby increasing the surface area of the food particles. Bile salts **emulsify** fat droplets, breaking them into smaller droplets so lipase can work faster.

Be careful not to say that bile salts break down fats. Make sure that you say 'fat droplets' otherwise it sound like bile salts are doing the same job as lipase.

Absorption

AQA	B3	✓
OCR B	B5	✓
EDEXCEL	B2	✓
WJEC	B2	✓
CCEA	B1	✓

KEY POINT

In the small intestine small digested food molecules are absorbed into the bloodstream by diffusion.

The inside of the small intestine is permeable and has a large surface area over which absorption can take place.

The lining of the small intestine contains two types of vessels that absorb the products of digestion:

- **Capillaries** absorb food and take it to the liver via the **hepatic portal vein**.
- **Lacteals** absorb mainly the products of fat digestion and empty them into the bloodstream.

A number of factors increase the surface area of the small intestine and so speed up the rate of absorption:

- The human small intestine is over five metres long.
- The inner lining is folded.
- The folds are covered with finger-like projections called **villi**.
- The villi are further covered by smaller projections called **microvilli**.

Other uses of digestive enzymes

AQA	B2	✓
OCR B	B6	✓
EDEXCEL	B3	✓

Microorganisms also make digestive enzymes. Decay organisms such as certain bacteria and fungi release these enzymes onto the food and take up the soluble products. These organisms are called **saprophytes** (saprotrophs).

Scientists have used microorganisms such as these to supply enzymes for a number of uses:

- Biological washing powders contain proteases and lipases.
- Proteases are used to pre-digest protein in some baby foods.
- **Amylases** are used to convert starch into sugar syrup.
- **Isomerase** is used to convert glucose into fructose, which is sweeter.

Microorganisms are also used to try and improve the health of the digestive system. **Probiotics** contain live bacteria that may produce useful vitamins and neutralise toxins. Prebiotics contain food substances that are said to encourage the growth of the 'good' bacteria.

PROGRESS CHECK

1. Why does bread start to taste sweet if it is chewed for several minutes?
2. What is the function of the gall bladder?
3. What are the products of fat digestion?
4. Why is fructose used in sweets rather than glucose?
5. The lining of the stomach is protected by mucus. Why does it need to be protected?
6. People who have coeliac disease may have many of their villi destroyed. What effect might this have on the process of absorption? Explain your answer.

6. Slows down the rate of absorption as there is a smaller surface area.
5. So that the acid does not damage it and the protease does not digest it.
4. It is sweeter so less is needed.
3. Fatty acids and glycerol.
2. To store bile.
1. The starch is being digested into the sugar maltose by amylase.

Sample GCSE questions

1 **(a)** The diagram shows an external view of the human heart shown from the front.

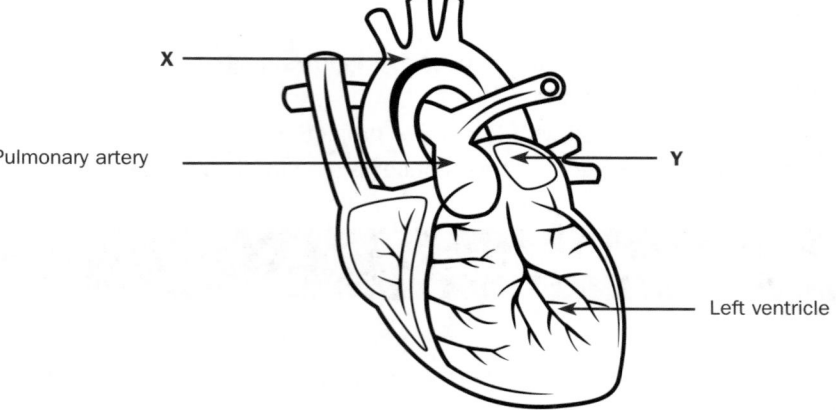

X

Pulmonary artery

Y

Left ventricle

(i) Identify structures X and Y on the diagram. [2]

X = aorta

Y = left atrium

> It is unusual to see a view of the outside of the heart. Most diagrams are sections of the heart which makes it easier. The two labels that are given should help you to label X and Y.

(ii) The pulmonary artery is labelled on the diagram.

It splits into two arteries. Why is this? [1]

The pulmonary artery has to carry blood to each lung.

(iii) Write down **two** differences in the composition of the blood found in the pulmonary artery and structure **X**. [2]

The blood in the pulmonary artery has less oxygen and more carbon dioxide.

> You could use the terms oxygenated or deoxygenated to describe how much oxygen is present in the blood.

(b) The heart has four chambers.

Explain why the walls of some of these chambers have different thicknesses.

The two atria have very thin walls because they only need to pump the blood down into the ventricles, which is not very far.

The right ventricle is thicker because it needs to pump blood to the lungs.

The left ventricle is the thickest because it needs to pump blood all around the body.

> This is a good answer but you should really include an explanation that the wall of the chambers contains muscle which generates pressure in the blood.

Sample GCSE questions

(c) The diagram shows a fish heart.

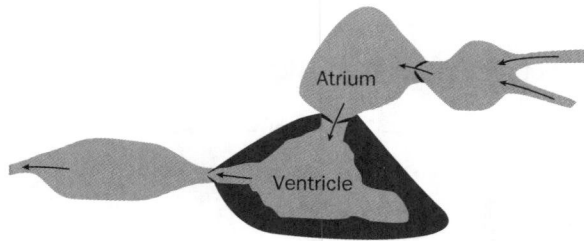

Fish only have a single circulation, not a double circulation.

Explain how you can tell this from this diagram. **[2]**

It only has one ventricle and one atrium.

This is because it does not have to pump blood to two different places.

> Having a double circulation also means that deoxygenated blood and oxygenated blood needs to be kept separate in the heart.

2 (a) The diagram shows a section through a human blood vessel with some red blood cells inside.

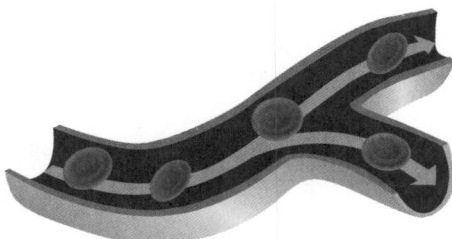

(i) What type of blood cell (artery, capillary or vein) is shown in the diagram? **[1]**

A capillary

(ii) Describe how the function of this blood vessel is related to its structure. **[2]**

The job of capillaries is to allow substances to pass between the blood and the tissues.

The wall of capillaries is very thin so substances can move in and out.

> The answer is correct but as well as saying that the wall is thin you should also say that it is only one cell thick.

(b) Describe **two** ways in which the shape or the structure of a red blood cell allows it to do its job efficiently. **[4]**

It is shaped like a biconcave disc. This gives it a large surface area compared to its volume so that oxygen can diffuse in and out quickly.

It also lacks a nucleus so that more haemoglobin can fit in to carry more oxygen.

> Red blood cells are also small and flexible so that they can squeeze through the smallest capillaries.

Exam practice questions

1 Rosie notices that if leaves are removed from a plant they dry out and shrivel.

She designs an experiment to investigate this water loss.

She picks three leaves from a plant and paints them with nail varnish as shown in the diagram.

Rosie then accurately measures the mass of each leaf and hangs them from a piece of string.

Nail varnish on top surface

Nail varnish on bottom surface

No nail varnish

Rosie reweighs the leaves at regular intervals.

Her results are shown on the graph.

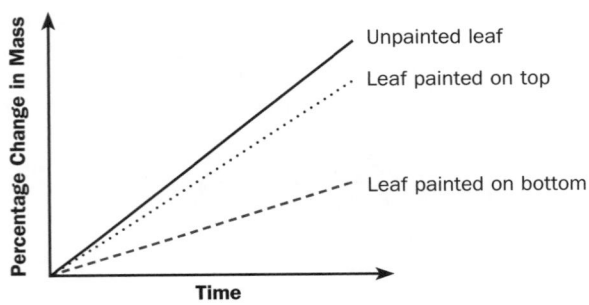

Unpainted leaf

Leaf painted on top

Leaf painted on bottom

(a) Describe what happens to the mass of the leaf with no nail varnish and explain the changes.

....

.... **[2]**

(b) Explain why the other two leaves change mass at different rates.

....

....

.... **[3]**

(c) The results of Rosie's experiment would have been different if she pointed a fan at the leaves. Explain why.

....

.... **[2]**

Exam practice questions

2   Until 1628, people thought that the blood was pumped by the heart.

They thought that the blood carried oxygen to the tissues and the blood then passed back to the heart in the same vessels.

(a)   Explain what was correct and what was incorrect with these ideas.

....

....

....   **[3]**

(b)   In 1628 William Harvey published the results of a series of experiments.

He tied a cord around the top of his arm.

He found that the veins below the cord became swollen with blood. Explain why this is.

....

....   **[1]**

(c)   Harvey then emptied the blood from part of the vein by rubbing his finger back in the direction O to H in this diagram.

He found that the blood was stopped from moving backwards by small structures in the vein.

(i)   What is the name of these small structures?

....   **[1]**

(ii)   Why are these small structures needed in veins but not in arteries?

....

....   **[1]**

Exam practice questions

(d) Harvey predicted that there must be small vessels joining arteries to veins but he could not see them.

What is the name of these vessels and suggest why Harvey could not see them.

....

....

.... **[2]**

3 The diagram shows the human digestive system.

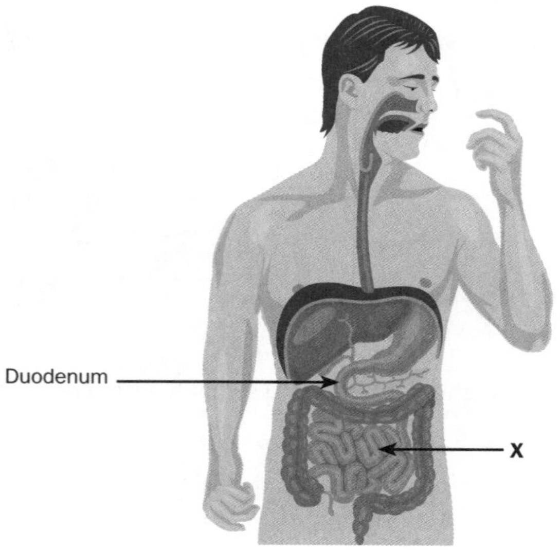

Duodenum

X

(a) **(i)** Write down the main function of the region labelled **X**.

.... **[1]**

(ii) Write down **two** ways in which the structure of **X** is adapted for this function.

1 ....

2 .... **[2]**

(b) One of the liquids added to the duodenum is bile.

Explain the functions of bile.

The quality of written communication will be assessed in your answer to this question.

....

....

.... **[4]**

8 Use, damage and repair

The following topics are covered in this chapter:

- The heart and circulation
- The skeleton and exercise
- The excretory system
- Breathing
- Damage and repair
- Transplants and donations

8.1 The heart and circulation

LEARNING SUMMARY

After studying this section, you should be able to:

- describe different types of circulatory systems
- describe the events and pressure changes in the cardiac cycle
- explain how the heart rate is controlled
- explain how blood clots and the nature of blood groups.

Different circulatory systems

OCR B   B5   ✓

Some organisms such as amoeba are small enough not to need a circulatory system.

Larger animals have different types of circulatory systems:

- Insects have an open circulatory system. The blood is not circulated in blood vessels, but moves around in the body cavity.
- Vertebrates have a closed circulatory system in which the blood is transported around the body in blood vessels. In fish this is a single system and the blood goes straight to the body from the gills. In humans there is a double system.

The structure of the heart and blood vessels and the advantages of a double circulatory system are covered on pages 137–138.

Our circulation has been investigated by many scientists throughout history:

'How Science Works' questions may ask for examples of how ideas in science have changed over the years. The discoveries made in the circulation are good examples to use.

Galen

Galen was doctor to five Roman emperors. He carried out dissections and showed that arteries carried blood, not air. He could not explain how the blood circulated.

Harvey

Harvey lived from 1578 to 1627. He carried out experiments and showed that the blood flows from the heart in arteries and back in veins. He could not see capillaries, but guessed that they were there.

The cardiac cycle

| OCR B | B5 | ✓ |
| WJEC | B3 | ✓ |

KEY POINT

The pattern of contraction of the different chambers of the heart is called the **cardiac cycle**:

- First the atria fill up with blood.
- They then contract and force the blood into the ventricles.
- The ventricles then contract and force blood out into the arteries.

The diagram shows how the pressure changes in the left atrium, left ventricle and aorta as these events happen.

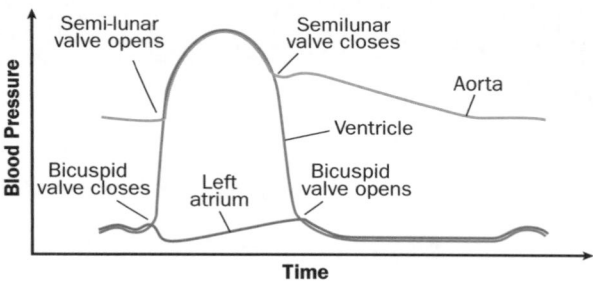

Pressure changes during the cardiac cycle.

As the blood flows through the blood vessels the pressure of the blood changes.

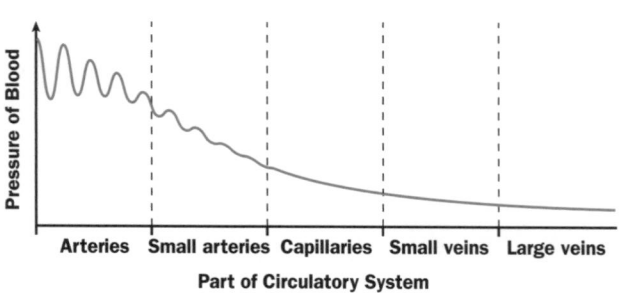

Pressure changes through the circulatory system.

The heart beat and pulse rate

OCR B   B5   ✓

The heart is made of powerful muscles which are supplied with food substances, including glucose and oxygen by the coronary artery. This constant supply of glucose and oxygen is needed for respiration to provide the energy for the contraction.

The pulse is a measure of the heart beat (muscle contraction) which puts the blood under pressure. This can be detected at various places, for example the wrist, ear and temple.

> **KEY POINT**
>
> The rate of the heart is controlled by small groups of cells called the **pacemaker**.

There are two areas called the **SAN (sinoatrial node)** and the **AVN (atrioventricular node)**. They produce a small electric current which spreads through the heart muscle making it contract. The rate of the pacemaker can be altered by hormones such as **adrenaline**.

The electric current that is produced by the pacemaker can be detected and studied using an **ECG** machine.

An ECG trace.

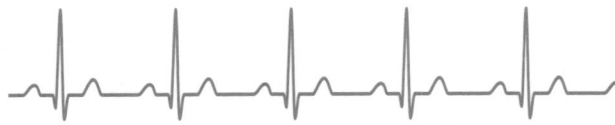

Remember that the heart will contract on its own, it does not need to receive nerve messages or hormones to make it contract. However, nerves and hormones can make the pacemaker speed up or slow down.

Blood groups and clotting

OCR B   B5   ✓
CCEA   B2   ✓

Different people have different blood groups and this is controlled by their genes. One of the main set of blood groups is the **ABO system**. Another is **rhesus** positive or negative. This describes the chemical groups or antigens found on the red blood cells.

The ABO system is determined by a single gene with three possible alleles A, B or O. A and B are codominant (fully expressed) and O is recessive to both. People with different blood groups will have different antibodies in their blood:

- People of group A have anti-B antibodies and those with group B have anti-A antibodies.
- People with group AB have neither antibodies.
- People of group O have both.

Blood clotting occurs when platelets are exposed to air, causing a series of chemical reactions leading to the formation of a mesh of fibrin fibres (clot). This can be affected by a number of factors:

- Haemophilia is an inherited condition in which the blood does not easily clot.
- Substances such as vitamin K, alcohol, green vegetables and cranberries affect clotting.
- Drugs such as warfarin, heparin and aspirin are used to reduce clotting.

PROGRESS CHECK

1. What type of circulation does a fish have?
2. Suggest why William Harvey could not show that capillaries existed.
3. What process in the heart muscle releases the energy for contraction?
4. Why are people sometimes given drugs such as heparin during and after an operation?
5. Compare the flow of blood in the arteries with that in the veins.
6. A person is blood group A and rhesus negative. Which antibodies do they have in their blood?

1. A single, closed circulation.
2. The microscopes at the time were not powerful enough.
3. Respiration.
4. To prevent blood clotting in blood vessels.
5. In the arteries, blood is under high pressure and flows in pulses rather than smoothly. It is under low pressure in veins.
6. Anti-B antibodies and rhesus antibodies, i.e. Anti-B and Rh+

8.2 The skeleton and exercise

LEARNING SUMMARY

After studying this section, you should be able to:

- recall the functions of the skeleton
- describe the structure of a synovial joint
- explain how the contraction of muscles can produce movement
- describe the components of a fitness programme

The skeleton

| OCR A | B7 | ✓ |
| OCR B | B5 | ✓ |

Different animals have different types of skeletons:

- Animals like insects have an **external** skeleton.
- All vertebrates have an **internal** skeleton. This makes it easier to grow and easier to attach muscles to.

The skeleton carries out important functions:

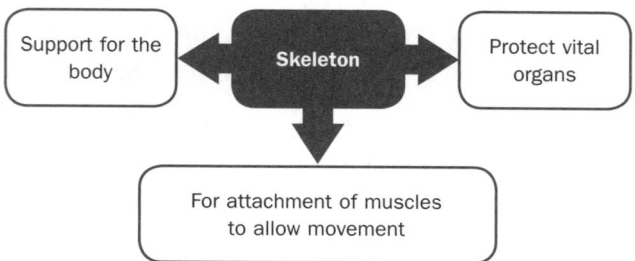

Support for the body ◄ **Skeleton** ► Protect vital organs

For attachment of muscles to allow movement

KEY POINT

The skeleton of vertebrates is made up of **bone** and **cartilage**, both of which are living tissues.

> **Remember** that our skeleton starts off as cartilage and turns to bone as our growing slows down. The amount of bone or ossification (the process of bone formation) can be used to tell how old a person is or was from their skeleton.

In animals like sharks the skeleton is mainly cartilage, but in humans it is mainly bone. The long bones of our arms and legs are hollow. This makes them much lighter, but still strong.

The structure of a long bone.

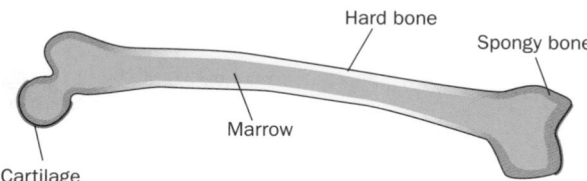

Hard bone

Spongy bone

Marrow

Cartilage

Joints

| OCR A | B7 | ✓ |
| OCR B | B5 | ✓ |

Where two bones meet they form a **joint**. Some of these joints are fused, but others allow movement.

> **KEY POINT**
>
> **Synovial joints** are specially adapted to allow smooth movement.

A synovial joint.

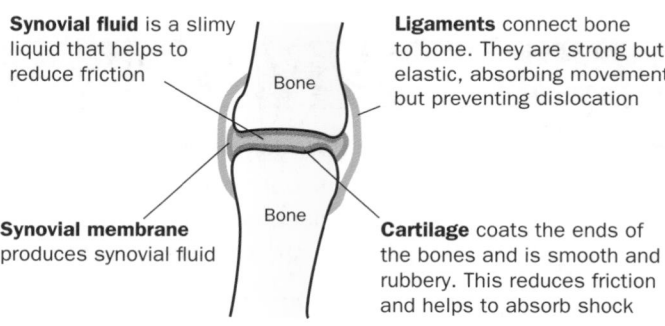

Synovial fluid is a slimy liquid that helps to reduce friction

Ligaments connect bone to bone. They are strong but elastic, absorbing movement but preventing dislocation

Bone

Bone

Synovial membrane produces synovial fluid

Cartilage coats the ends of the bones and is smooth and rubbery. This reduces friction and helps to absorb shock

This joint is a **hinge joint** such as the joints between the fingers and the joint at the elbow or knee. These joints allow movement in one dimension.

Ball and socket joints, such as the hip joint, have almost all round movement.

Muscles are the main effectors in the body. They contain muscle fibres that can shorten and so make the muscle contract. In order to occur, this needs energy from respiration. Muscles can only contract and pull on bones, but cannot actively expand or push.

> **Remember** that people who are described as being antagonistic, are always arguing and taking the opposite point of view. In the same way antagonistic muscles work against each other.

Muscles therefore have to be arranged in **antagonistic pairs**. When one contracts the other relaxes and vice versa.

> **KEY POINT**
>
> Muscles are connected to bones by **tendons**.

Tendons are very strong and are not elastic.

Muscles and levers

OCR B   B5   ✓

The biceps and triceps are antagonistic muscles in the human arm.

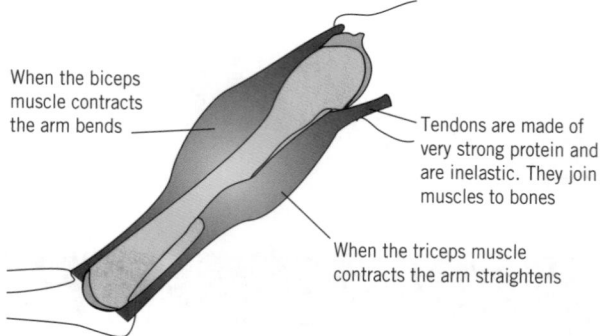

When the biceps muscle contracts the arm bends

Tendons are made of very strong protein and are inelastic. They join muscles to bones

When the triceps muscle contracts the arm straightens

> **KEY POINT**
>
> The arm works like a **lever**, with the elbow being the **pivot**.

The muscles are attached close to the pivot so this means that:

- A larger distance is moved by the hand than the muscles.
- A larger force is exerted by the muscles than is exerted by the hand.

Exercise and fitness

OCR A   B7   ✓

Many people now visit a gym and regular exercise can improve a person's fitness. It can also improve their general health, helping to prevent diseases such as heart disease. When a person decides to undertake a proper fitness programme they should follow these steps:

> First discuss factors in the person's lifestyle, such as alcohol and tobacco consumption, family medical history and any medication they are taking.

→

> Then devise a fitness programme weighing up any side effects with any benefits that may be gained.

↓

OCR B candidates need to know that there are different ways of measuring fitness and should be able to describe some of the methods.

> Modify the programme if the fitness is improving faster than expected or in the case of an injury.

←

> Monitor the progress of the training including changes in heart rate, blood pressure and recovery period.

↓

> Decide how successful the treatment has been, taking into account the accuracy of the monitoring technique used and the repeatability of the data.

PROGRESS CHECK

1. Why are synovial joints called synovial?
2. Why do the bones of birds need to be particularly light?
3. What type of joint is found in the hip?
4. What are the differences between ligaments and tendons?
5. The triceps muscle is called an extensor muscle. Why is this?
6. What is the 'recovery period' that is measured when exercising?

1. Because they contain synovial fluid to allow smooth movement.
2. So that the bird can fly easily using as little energy as possible.
3. A ball and socket joint.
4. Tendons join muscle to bone and ligaments join bone to bone and are more elastic.
5. When it contracts, the arm is extended or straightened.
6. The recovery period is the time it takes for the heart rate to return to normal after exercise has finished.

8.3 The excretory system

LEARNING SUMMARY	**After studying this section, you should be able to:**
	• recall the main waste products of the body
	• describe the structure and function of the kidney
	• describe how urea is made and where it is excreted.

Different waste materials

AQA	B3	✓
OCR B	B5	✓
EDEXCEL	B3	✓
WJEC	B3	✓

The body produces different types of waste as a result of its metabolism. Many of these wastes are toxic and so must be removed from the body. The removal of these wastes is called **excretion**.

The sites of production of excretory products.

Remember that in excretion the waste product has to be made by the body. Most of faeces are not made of excretory products. It has been taken in, passed through the gut and passed out. This is called egestion.

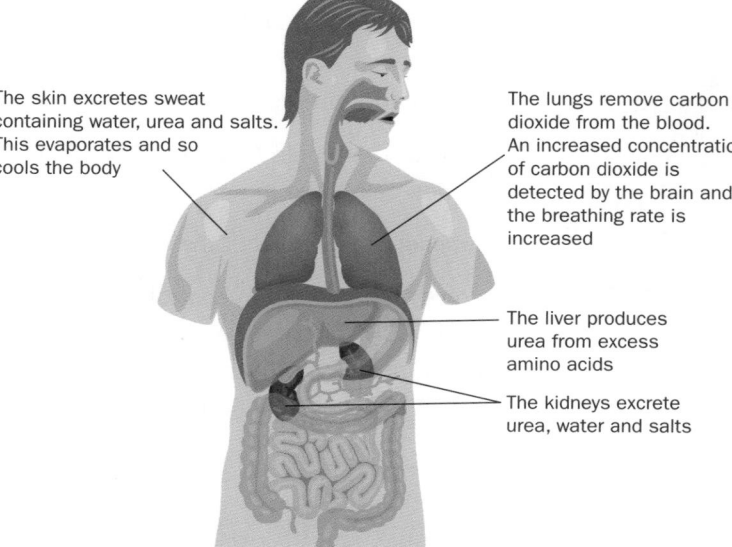

The skin excretes sweat containing water, urea and salts. This evaporates and so cools the body

The lungs remove carbon dioxide from the blood. An increased concentration of carbon dioxide is detected by the brain and the breathing rate is increased

The liver produces urea from excess amino acids

The kidneys excrete urea, water and salts

How the kidneys work

AQA	B3	✓
OCR B	B5	✓
EDEXCEL	B3	✓
WJEC	B3	✓

The kidneys produce urine by the following process:

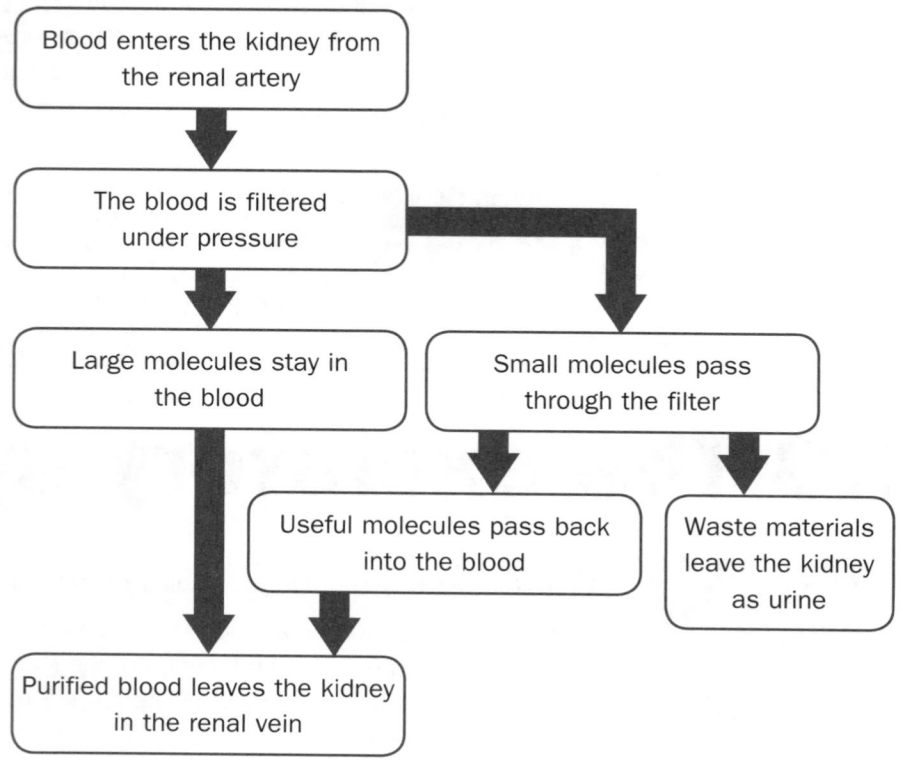

It is important to realise that although there is no glucose in the urine and also no protein, the reasons for this are different in each case.

The useful substances that pass back into the blood are:

- All the sugar and amino acids (re-absorbed by active transport).
- The dissolved minerals needed by the body.
- As much water as is needed by the body.

The kidneys are made of millions of small tubes called **kidney tubules**.

> **KEY POINT**
>
> Each individual kidney tubule is called a **nephron**.

Remember that selective re-absorption means that some substances (like glucose and amino acids) are re-absorbed and others are not. As this is against a concentration gradient selective re-absorption must use active transport.

Different parts of these tubules do different jobs in the production of urine.

The part of the tubule that filters the blood is made of a tight knot of blood capillaries called the **glomerulus**.

The **capsule** then collects the fluid that is forced out of the glomerulus.

The job of the kidneys in controlling the water balance of the body involves the hormone ADH. This is described on page 15. ADH acts on the final part of the nephron, making it more permeable to water. More water is therefore reabsorbed and more concentrated urine is produced.

Structure of a kidney tubule.

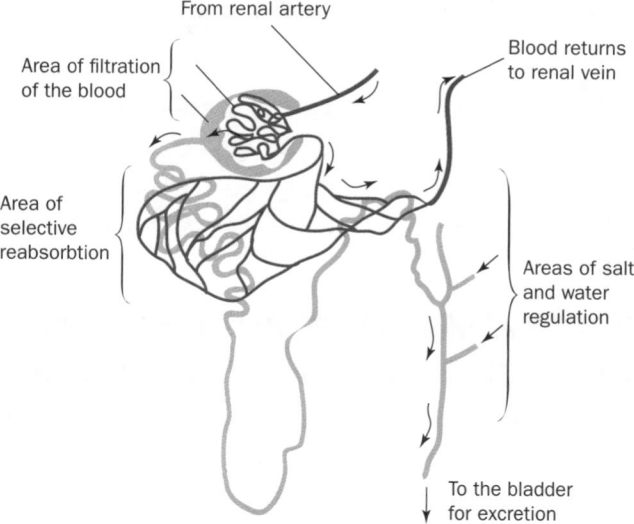

From renal artery

Area of filtration of the blood

Blood returns to renal vein

Area of selective reabsorbtion

Areas of salt and water regulation

To the bladder for excretion

Urea production

| OCR B | B5 | ✓ |
| EDEXCEL | B3 | ✓ |

When the body takes in proteins they are digested into amino acids and absorbed into the bloodstream. The body cannot store amino acids so if there are too many then they are either built up into proteins or destroyed.

KEY POINT

The liver breaks down excess amino acids releasing urea.

Urea is sent to the kidneys for excretion.

PROGRESS CHECK

1. What is contained in urine?
2. If the blood pressure drops, the kidneys stop working. Why is this?
3. Why is there usually no protein in the urine?
4. Why are there usually no amino acids in the urine?
5. Alcohol reduces ADH production. What effects can this have?
6. Why is there more urea in a person's urine the day after eating a high protein meal?

1. Water, salts and urea.
2. A certain blood pressure is needed to filter the blood.
3. Protein molecules are too large to get through the filter.
4. They are filtered out of the blood but then are reabsorbed.
5. Less water is reabsorbed so a greater volume of urine is produced.
6. The protein is digested into amino acids and the excess amino acids are broken down into urea which is excreted in the urine. Urine is more concentrated.

8.4 Breathing

LEARNING SUMMARY

After studying this section, you should be able to:

- describe the process of gaseous exchange in different animals
- explain the process of breathing in humans
- interpret breathing patterns from spirometer traces.

Gaseous exchange in different animals

OCR B   B5   ✓

Different types of organisms have different systems for obtaining oxygen and losing carbon dioxide.

> **KEY POINT**
>
> The moving of these gases between the organism and the environment is called **gaseous exchange**.

In small organisms, such as worms, gaseous exchange occurs over the whole body surface.

Fish have **gills** for gaseous exchange. The filaments take in oxygen from the water.

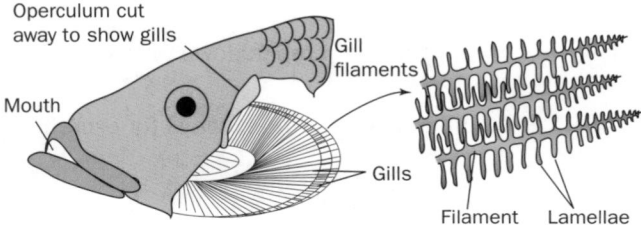

Operculum cut away to show gills · Mouth · Gill filaments · Gills · Filament · Lamellae

Gill structure.

The presence of gills means that fish must live in water. Amphibians such as frogs have lungs, but also get oxygen through their skin. This needs to be moist so they can live on land, but only in damp places.

Humans have **lungs** for gaseous exchange.

> **KEY POINT**
>
> The gaseous exchange takes place in millions of tiny air sacs called **alveoli**.

Adaptations for gaseous exchange

AQA   B3   ✓
OCR B   B5   ✓
WJEC   B2   ✓
CCEA   B1   ✓

The human alveoli are adapted for efficient gaseous exchange in a number of ways:

- They are permeable.
- They have a moist surface.
- There are millions of alveoli providing a surface area of about 90 m^2.
- There are many blood vessels providing a rich blood supply.

Fish gills allow efficient gaseous exchange in water because the filaments:

- have a large surface area
- are very thin
- are well supplied with blood.

Breathing mechanisms

AQA	B3	✓
OCR B	B5	✓
WJEC	B2	✓
CCEA	B1	✓

Large and active organisms cannot simply rely on diffusion to bring enough oxygen to their respiratory surfaces.

> **KEY POINT**
>
> Organisms like mammals and fish **ventilate** their respiratory surfaces using muscular contractions. This is called **breathing**.

In humans, drawing air in and out of the lungs involves changes in pressure and volume in the chest. These changes are brought about by contractions of the diaphragm and intercostal muscles. They work because the pleural membranes form an airtight pleural cavity.

Breathing in (inhaling):

- The intercostal muscles contract moving the ribs upwards and outwards.
- The diaphragm contracts and flattens.
- Both of these actions will increase the volume in the pleural cavity and so decrease the pressure.
- Air is drawn into the lungs because the air moves from the higher atmospheric pressure into the lungs where there is lower pressure.

Breathing out (exhaling):

- The intercostal muscles relax and the ribs move down and inwards.
- The diaphragm relaxes and domes upwards.
- The volume in the pleural cavity is decreased so the pressure is increased.
- Air is forced out of the lungs.

> Make sure that you know the difference between breathing and respiration. Many candidates confuse the two. Breathing is muscular movements that aid gaseous exchange, but respiration is the reaction that releases energy from food.

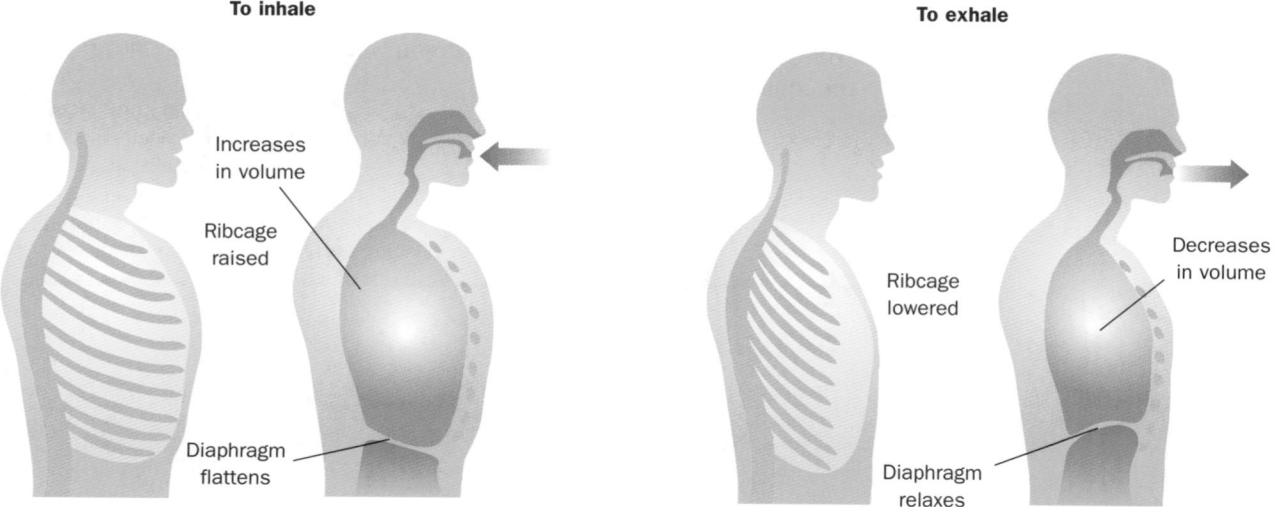

To inhale

Increases in volume

Ribcage raised

Diaphragm flattens

To exhale

Decreases in volume

Ribcage lowered

Diaphragm relaxes

In fish, water enters mouth and the mouth then closes and forces water out across the gills.

Measuring lung volumes

OCR B   B5   ✓

Looking at spirometer traces like this can often give indications of disorders such as asthma (see page 162).

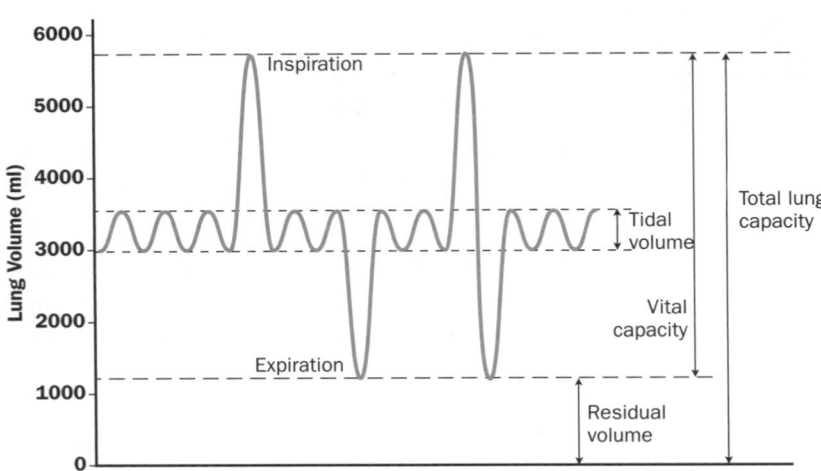

A spirometer trace.

- **Total lung capacity** is the maximum volume of air that the lungs can hold.
- **Vital capacity** is the maximum volume of air that the lungs can pass in or out per breath.
- **Residual volume** is the air that is left in the lungs after a person has breathed out deeply.
- **Tidal volume** is the volume of air exchanged per break at rest.

PROGRESS CHECK

1. How does a worm obtain oxygen?
2. Why do so many frogs live in tropical rainforests?
3. Where in fish gills does oxygen enter the bloodstream?
4. What happens to the diaphragm when we breathe in?
5. Why do gill filaments look pink?
6. What is the vital capacity of the person whose breathing is shown on the spirometer trace?

6. About 4500 ml.
5. Because they have a rich blood supply.
4. It contracts and flattens.
3. In the filaments.
2. Because it is moist enough for them to breathe through their skin.
1. By diffusion over its body surface, through its skin.

8.5 Damage and repair

After studying this section, you should be able to:

- describe the main problems that can occur with the heart, kidney, lungs and joints
- explain how a kidney dialysis machine works.

Heart problems

OCR B   B5   ✓

There are many heart conditions and diseases that might affect the working of the heart. These include:

Condition	Possible effect on the body	Possible treatment
An irregular heart beat	Less oxygen in the blood	Drug treatment
A hole in the heart	Blood can move directly from the right side to the left side of the heart	Surgery
Damaged or weak valves	Reduces blood circulation	Replacement by artificial valves
Coronary heart disease	Reduces blood flow to the heart muscle and can lead to heart attacks	By-pass surgery

If the damage is bad enough the person may need a heart transplant.

Kidney disease

AQA   B3   ✓
OCR B   B5   ✓
EDEXCEL   B3   ✓
WJEC   B3   ✓

People may have kidney failure for a number of reasons. A person can survive if half of their kidney tubules are still working, but if the situation worsens there are two options:

Kidney dialysis ← Kidney failure → Kidney transplant

Kidney dialysis involves linking the person up to a dialysis machine. This takes over the job of the kidneys and removes waste substances from the blood.

A dialysis machine.

Remember that the dialysis machine works using diffusion. Candidates often think that it involves osmosis because the cellophane is partially permeable.

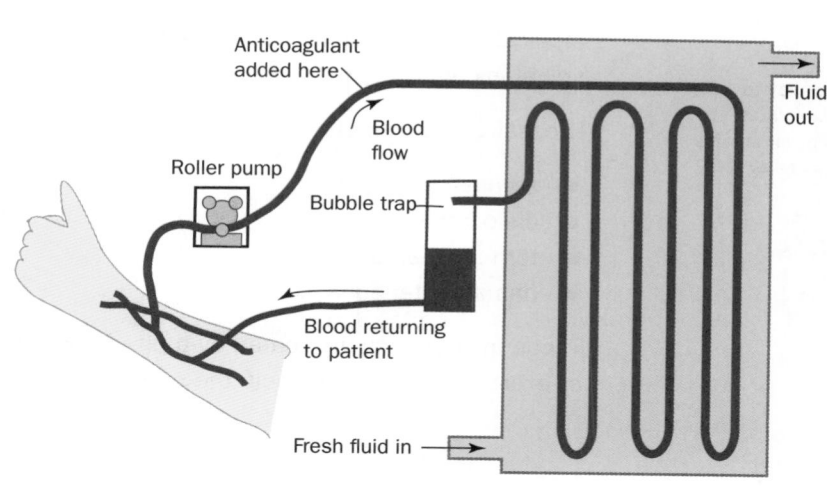

The blood is removed from a vein in the arm. It then passes through a long coiled tube made of partially permeable cellophane. The fluid surrounding the tube contains water, salts, glucose and amino acids, but no waste materials, such as urea. These waste materials therefore diffuse out of the blood into the fluid.

Lung problems

| OCR B | B5 | ✓ |
| WJEC | B2 | ✓ |

The lungs are easily infected because they are a 'dead end'. Microbes and particles can easily collect there. Some of these types of problems are:

- **Industrial diseases** such as asbestosis where small particles of asbestos are breathed in and damage the lungs.
- **Genetic conditions** such as cystic fibrosis where too much mucus is made.
- **Lifestyle factors** such as smoking which can lead to lung cancer. Some of the effects of smoking on the lungs are described on page 45.

The respiratory system tries to protect itself from disease by producing mucus and the action of cilia. These are shown on page 35.

More and more people now have asthma. The cause of this is not yet understood. The symptoms of asthma are difficulty breathing, wheezing and a tight chest. It is treated using inhalers. If not treated, asthma can be fatal. During an asthma attack:

- The lining of the airways become inflamed.
- Fluid builds up in the airways.
- The muscles around the bronchioles contract, constricting the airways.

Fractures and sprains

| OCR A | B7 | ✓ |
| OCR B | B5 | ✓ |

Despite being strong, bones can still be broken by a sharp knock. This is more likely to happen to elderly people because they often have **osteoporosis** which makes their bones weaker. There are three types of bone fractures:

- A **simple** fracture of the bone is a fracture without the skin being broken.
- A **compound** fracture of the bone is where the skin has been broken.
- A **green stick** fracture is where the bone has not been fractured all the way through.

If a person has a fracture they should not be moved, other than by trained medical professionals, as this could cause further damage (especially if it is a spinal injury).

Excessive exercise can cause damage to other parts of the body. These include:

The best way to remember the treatment for sprains is to remember RICE:

<u>R</u>est
<u>I</u>ce
<u>C</u>ompression
<u>E</u>levation.

- sprains
- dislocations
- torn ligaments
- damaged tendons.

These injuries may be treated by a physiotherapist who will devise a suitable set of exercises that should help the person recover.

8.6 Transplants and donations

LEARNING SUMMARY	**After studying this section, you should be able to:**
	• recall the main parts of the body that can be transplanted
	• discuss the ethics of transplants
	• analyse blood groups to decide on successful matches for transfusions.

Transplants

OCR B	B5	✓
WJEC	B3	✓

People are living longer now than they ever have been. This is due to less industrial disease, healthier diet and lifestyle, modern treatments and cures for disease and better housing. This means that there is an increasing demand for transplants of different body parts:

- Kidney – as a person can survive with one kidney it is possible for a person to donate one kidney to be transplanted into another person. Other transplants may come from dead donors.
- Heart – sometimes the heart is too badly damaged and surgery cannot repair the problem.
- Lung – the heart is often transplanted along with the lungs in a heart-lung transplant.
- Joints – operations to replace or resurface the ends of bones are now common.

The main problem with transplants is to prevent the person's immune system rejecting the transplanted organ. This is avoided by taking certain precautions:

- Making sure that the donor has a similar 'tissue type' to the patient.
- Treating the patient with drugs for the rest of their life to make their immune system less effective.

Some replacements are mechanical rather than biological. The ends of bones may be capped with metal and heart assisted pumps may be inserted. Some of the problems of using mechanical replacements include their size, using materials that will not react with the body and in some cases, the need for a power supply.

The ethics of transplants

OCR B	B5	✓
EDEXCEL	B3	✓
WJEC	B3	✓

The increasing demand for transplants has led to a major problem, a serious shortage of donors. For some people this means waiting for a long time and some may die before a suitable organ becomes available. The line on the graph shows the number of people in the UK on the transplant list. The bars show the number of people who died and became donors and the number of transplants performed.

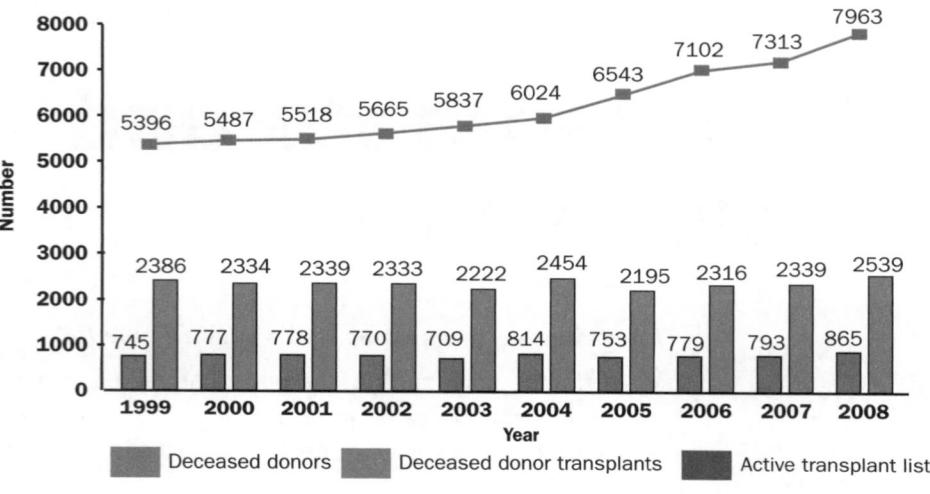

The current system only allows organs to be removed if a person carries a donor card. Relatives of the person can stop the organs being taken even if the person is carrying a card. A new system has been suggested which would mean that organs can be taken from a person unless they have opted not to allow it. This is called **presumed consent**. This would greatly increase the numbers of organs available for transplants, but not everyone is happy with the idea. There are people who have religious and cultural objections. They might not realise that they need to opt out and could have their organs used. Mistakes could also be made and people could have their organs taken by mistake even if they have opted out. Getting it wrong could lead to distress for relatives and could lead to a backlash against doctors and organ donation.

> 'How Science Works' questions often ask for arguments for and against a new scientific development. Make sure that you can give both sides of this argument.

Blood transfusions

| OCR B | B5 | ✓ |
| CCEA | B2 | ✓ |

When giving blood transfusions it is important to make sure that the blood groups are matched. Blood groups are discussed on page 151. Unsuccessful blood transfusions cause **agglutination** (blood clumping). The antigens on red blood cells and the antibodies in the blood serum will decide how blood groups react and therefore if the blood transfusion is successful. The table shows which blood groups are compatible.

	Blood group of donor							
Type	O–	O+	B–	B+	A–	A+	AB–	AB+
AB+	■	■	■	■	■	■	■	■
AB–	■	□	■	□	■	□	■	□
A+	■	■	□	□	■	■	□	□
A–	■	□	□	□	■	□	□	□
B+	■	■	■	■	□	□	□	□
B–	■	□	■	□	□	□	□	□
O+	■	■	□	□	□	□	□	□
O–	■	□	□	□	□	□	□	□

(The left-hand vertical label reads "Blood group of receiver")

■ = successful transfusion

Gamete donations

OCR B   B5   ✓

Some types of infertility can be treated with hormones. This is covered on page 18. In other cases sperm or egg production is not possible. These cases can be treated by:

● **Egg donation** which would involve IVF or ovary transplants.
● **Artificial insemination** which sometimes involves donated sperm, although often the sperm is from the biological father.

Sometimes the woman's body cannot support a baby throughout pregnancy and another female may have a baby for her. This is called **surrogacy**.

PROGRESS CHECK

1. Why are kidney transplants between relatives more successful?
2. Write down one problem with having a battery powered heart assist pump inserted into the body.
3. Write down one property needed in the metal that is used to replace bone in joints.
4. Suggest one possible problem with surrogacy arrangements.
5. A person who is blood group A- cannot successfully donate blood to a B- person. Explain why.
6. People with blood group O- are usually described as universal donors. Why is this?

1. Genetically more similar so less likely that rejection will occur.
2. The battery might need to be changed.
3. Light/does not corrode/smooth/strong.
4. The surrogate may not want to hand over the baby. Could accept that the baby is not genetically linked to the mother.
5. People with B- blood have anti-A antibodies in their blood and so would attack the red blood cells causing agglutination.
6. They can give their blood to people of any other blood group as their blood contains no anti-A or B antigens.

Sample GCSE questions

1 **(a)** The graphs show changes in volume inside the chest cavity during one complete breath.

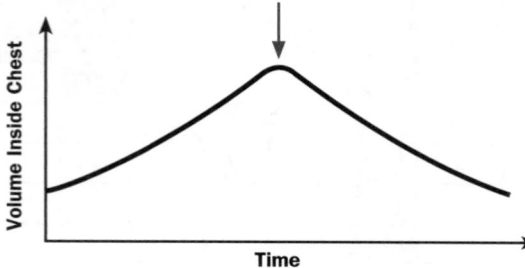

(i) On the graph draw an arrow to show when the person starts to breathe out. **[1]**

(ii) Explain how the changes in the volume of the chest are brought about during **breathing in**. **[3]**

> The intercostal muscles contract moving the ribs upwards and outwards.
>
> Also the diaphragm contracts and flattens.
>
> Both of these processes increase the volume in the chest.

It is important to realise that the changes in the intercostal muscles and diaphragm increase the volume of the lungs and not the other way round.

(b) The diagram shows part of the system of tubes leading into some air sacs in the lungs.

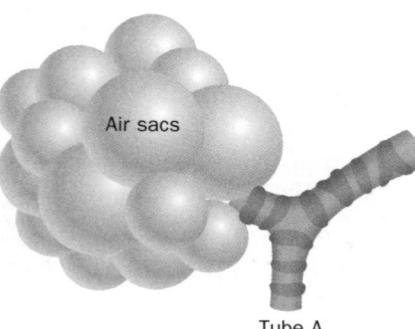

Air sacs

Tube A

(i) Write down the name of the small tube A that leads into the air sacs.

> bronchioles

(ii) Use the diagram to help explain one way that the lungs are adapted for gaseous exchange. **[2]**

> There are large numbers of air sacs in the lungs. This provides a large surface area for gaseous exchange.

There are many adaptations shown by the lungs but this is the one seen in the diagram.

Sample GCSE questions

(iii) Explain how changes to the structures shown on the diagram may cause an asthma attack. **[2]**

> Muscle in the bronchi and bronchioles contracts. This causes the diameter to decrease making it harder to draw air into the lungs.

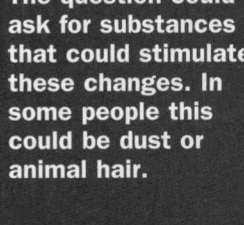

The question could ask for substances that could stimulate these changes. In some people this could be dust or animal hair.

(c) The graph shows the volume of air breathed out by two different people during one breath. Both people are the same size.

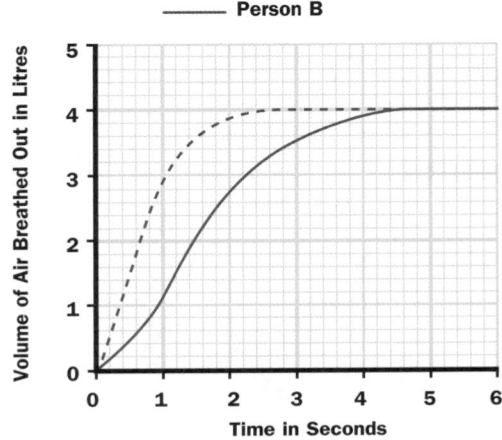

– – – – Person A

———— Person B

(i) Compare the pattern of breathing shown by these two people. **[3]**

> Both people breathe out the same volume of air.
>
> This is four litres.
>
> Person B takes longer to finish breathing out this amount. Person B takes 4.6 seconds but person A only takes 2.6 seconds.

A good answer. It is important to quote data from graphs to back up your answer. You could say that they both have the same vital capacity.

(ii) Which person is most likely to have asthma? Explain your answer. **[1]**

> Person B, because their bronchi have constricted so it is harder to draw air in.

(iii) Why is it important to know that these two people are the same size when comparing their breathing? **[1]**

> Because larger people tend to have larger lungs so it would not be a fair comparison.

A good answer. Some candidates might just say 'to make it a valid comparison' but this answer explains why.

Exam practice questions

1 Rebekah is given four different solutions by her teacher.

The teacher says that they are similar to liquids that could be taken from different parts inside the kidney.

Solution A is similar to urine.

Solution B is similar to the filtrate produced by the filter unit.

Solution C is similar to plasma from the renal vein.

Solution D is similar to plasma from the renal artery.

(a) The diagram shows a kidney tubule (nephron) and blood vessels.

Use label lines and the letters **A**, **B**, **C** and **D** on the diagram to show the position from which each of the four solutions could be taken. **[4]**

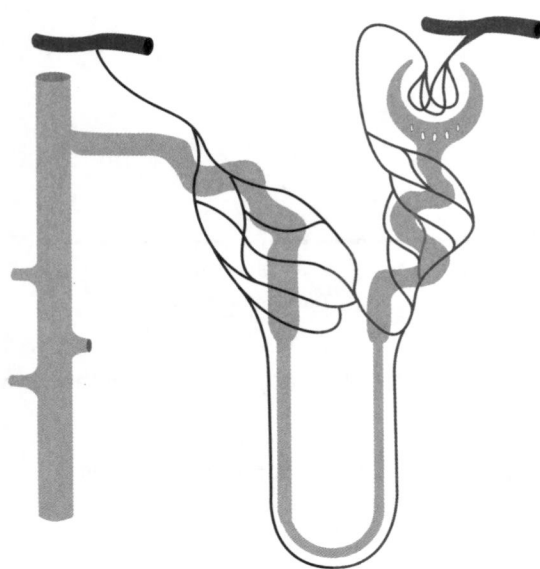

(b) Rebekah tests each of the solutions to see if they contain glucose or protein.

Her results are shown in the table.

Solution	Results of each test	
	Glucose	Protein
A	absent	absent
B	present	absent
C	small amount present	present
D	present	present

(i) Explain why solution A (urine) did not contain any protein.

....

.... **[1]**

Exam practice questions

(ii)   Explain why solution A (urine) did not contain any glucose.

....

....   **[2]**

(iii)   Rebekah found that there was less glucose in solution C (renal vein) than in solution D (renal artery). Explain why.

....

....   **[2]**

(c)   Rebekah's teacher tested the solutions for urea.

He found that there was more urea in solution D (renal artery) than in solution C (renal vein).

Explain this difference.

....

....   **[2]**

2   The diagram shows a kidney dialysis machine.

(a)   Why is the clot and bubble trap important?

....

....   **[2]**

Exam practice questions

(b)   Use the diagram to explain how the kidney dialysis machine removes waste materials from the blood.

....

....

....   **[2]**

(c)   What must the fresh dialysis fluid contain? Explain why.

....

....

....   **[2]**

3   The diagram shows the main bones and muscles of the human arm.

(a)   **(i)**   Label the biceps muscle on the diagram.   **[1]**

  (ii)   What happens to the arm when this muscle contracts?

....   **[1]**

  (iii)   How is this muscle connected to the bones of the arm?

....   **[1]**

  (iv)   The biceps and the triceps are called antagonistic muscles.

Why is this?

....

....   **[2]**

(b)   **(i)**   Write down the type of joint found at the elbow.

....   **[1]**

Exam practice questions

(ii) What are the functions of cartilage in the joint?

....

.... **[2]**

(c) Some people have problems with the joint in their elbow.

They can have the joint replaced by an artificial joint.

Suggest what properties are needed in the material used to make artificial joints.

....

.... **[2]**

(d) The table shows some data about the number of patients receiving an artificial joint in one hospital.

It also shows the number of patients having problems as a result of the operation.

Age of patients	Younger than 55	56 to 70	Older than 70
Number having operation	36	42	22
Number having problems	16	28	15

(i) Suggest explanations for the differences between the numbers of people having the operation at different ages.

....

....

.... **[2]**

(ii) The percentage of 56 to 70 year olds having problems is 66.7%.

What is the percentage of people younger than 55 who have complications?

answer = ... % **[2]**

(iii) Suggest a reason for the difference between these two figures.

.... **[1]**

9 Animal behaviour

The following topics are covered in this chapter:

- **Types of behaviour**
- **Communication and mating**

9.1 Types of behaviour

LEARNING SUMMARY

After studying this section, you should be able to:

- decide whether a pattern of behaviour is innate or learned
- describe the two types of conditioning
- describe the nature of memory
- describe examples of evidence for the development of the human brain.

Innate or learned behaviour

OCR A   B6   ✓
EDEXCEL B3   ✓

> **KEY POINT**
>
> When animals are born they have certain inbuilt types of behaviour. This is called **innate behaviour**.

It includes simple reflexes and **instincts**. Instincts are much more complicated actions than reflexes. Innate behaviour is controlled by the genes and is inherited from the animal's parents. It is important so that the young animal has certain skills needed to survive before it has the chance to learn. The scientist Nikolaas Tinbergen studied seagulls to see what innate behaviour the gulls used to help their young to collect food from their parents.

> **KEY POINT**
>
> **Learning** is a change in behaviour caused by experiences.

There are several types of learned behaviour:

- Probably the first learning experience of animals is **imprinting**. New born animals will be attracted to the first animal or object that they see. The first scientific studies of this were carried out by Konrad Lorenz. He discovered that if geese were reared by him from hatching, they would treat him like a parent bird and follow him around.
- Another way that animals learn is by **habituation**. An animal may be frightened by a particular stimulus such as a loud noise. However, if the stimulus is repeated and the animal is not harmed then the animal learns

to ignore the stimulus. It has been habituated. This idea is often used in the training of animals such as police horses.

- Animals can also learn by **conditioning**. Instead of learning to ignore a stimulus they will associate one stimulus with another. This was first discovered by a Russian called Ivan Pavlov. This type of learning is also used in training animals for specific jobs.

- **Insight** learning can be shown by apes in which they use their intelligence to solve a problem.

Types of conditioning

| OCR A | B6 | ✓ |
| EDEXCEL | B3 | ✓ |

There are two main types of conditioning:

- **Classical conditioning** was the type shown by Pavlov's dogs. Pavlov began to ring a bell each time the dog was shown their food. After a while Pavlov found that the dogs salivated when the bell was rung regardless of whether food was present. The dog had become conditioned so that it associated a bell with the arrival of food. The response is caused by a stimulus different from the one that originally triggered it.

- **Operant conditioning** is sometimes called 'trial and reward learning'. It might involve giving animals (for example a rat) a food reward if they press a lever after seeing a light or hearing a sound. Some sort of punishment can also be used as a negative reinforcement. Unlike classical conditioning (e.g. pressing a lever), the response is not natural behaviour. Dogs producing saliva is natural, but rats pushing levers is not.

> Use ideas about conditioning to reward yourself when you finish revising a topic and do well in a test!

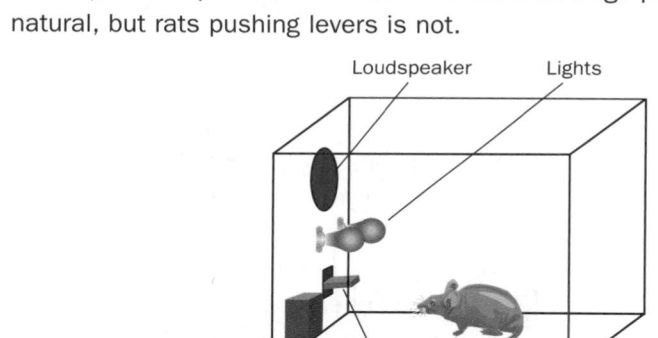

Loudspeaker   Lights

Food dispenser   Response lever

Memory and the brain

| OCR A | B6 | ✓ |
| EDEXCEL | B3 | ✓ |

In order to learn, organisms must be able to remember past events.

> **KEY POINT**
>
> **Memory** is the storing of information in the brain so that it can be retrieved.

> The location of different functions in different parts of the brain can be investigated by electrical stimulation and with brain scans such as MRI.

This takes place in a part of the brain called the **cerebral cortex**. This part of the brain is also responsible for intelligence, language and consciousness. When we learn something certain pathways in the brain are formed and others are lost. The more we repeat the task, the more likely that the neural pathway will stay connected. Verbal memory can be divided into short-term and long-term memory. It seems that some parts of the brain may only be able to learn skills up to a certain age. For example, children who have been brought up away from other people have difficulty learning language.

Human development

EDEXCEL B3 ✓

The development of the human brain in evolution has provided humans with a greater ability to learn. This has enabled a better chance of survival. The evidence for human evolution has come from a number of sources:

- The discovery of ancient **fossils** have provided much evidence, especially skulls. A scientist called Louis Leakey discovered human fossils in Africa dated from about 1.6 million years ago. A 3.2 million year old fossil nicknamed Lucy has been found and recently a 4.4 million year-old fossil called Ardi.

Fossil skulls of human ancestors.

 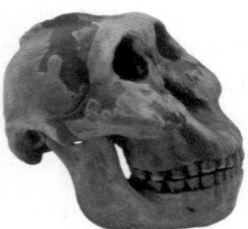 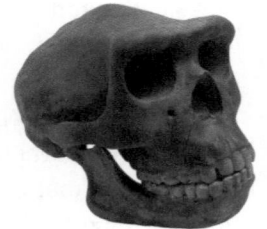 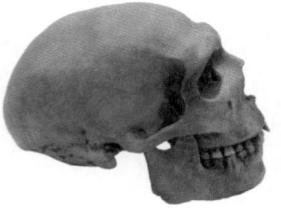

> Remember that mitochondrial DNA is not inherited in the normal way, i.e. half from our mother and half from our father. Our mitochondrial DNA comes entirely from our mother.

- Discovery of ancient tools can tell us a lot about ancient human behaviour and skills.
- The study of a special type of DNA, which is found in mitochondria, has suggested that humans evolved in Africa.

PROGRESS CHECK

1. Why is imprinting important for young animals?
2. Why are police horses exposed to loud noises during their training?
3. Why do scarecrows stop having an effect on birds after a while?
4. Which part of our brain contains our memories?
5. How could operant conditioning be used to teach a dog a trick?
6. Why can studying fossil skulls give us information on the intelligence of ancient humans?

1. So that they stay with their mother for food/protection/to learn and to remain safe.
2. So that they become habituated to loud noises and do not respond to them when working.
3. The birds become habituated and are not scared – it becomes unassociated to danger.
4. The cerebrum/cerebral cortex.
5. Give them a reward every time they do the trick.
6. The size of the brain that they contained can be worked out which gives an idea of intelligence.

9.2 Communication and mating

LEARNING SUMMARY

After studying this section, you should be able to:

- describe the function of courtship
- explain the function of parental care
- compare communication methods in different organisms
- describe some of the studies of great ape behaviour in the wild.

Finding a mate

EDEXCEL B3 ✓

Sexual reproduction involves individuals choosing who to mate with. This is particularly important for females as they use energy and food reserves producing the young.

> **KEY POINT**
>
> **Courtship** is a type of behaviour used to help to choose a mate.

In many types of animals, males show courtship behaviour to try and persuade females that they have good genes and can provide for the offspring. This is why in many species the male is more brightly coloured than the female. Once a mate is chosen there are a number of different types of relationship:

- Very few animals mate with the same partner for life. The albatross is an exception. Albatrosses can live to be 80–85 years old and they mate for life.
- Some animals live in groups with a dominant male that mates with all the females.
- Other animals may have different mates each year or even in the same breeding season.

Parental care

EDEXCEL B3 ✓

> **KEY POINT**
>
> Most birds and mammals look after their young for some time after they are born. This is called **parental care**.

This means that these animals have developed special feeding behaviour. Baby birds will call to their parents and often gape with their mouth wide open.

Often the inside of the baby's mouth is coloured to attract attention of the parent bird.

Mammals feed their young on breast milk. The young are born with an instinct to suck on the breast. The process of breast feeding is also important in building an emotional bond between the mother and the baby. Showing parental care to their babies has advantages and disadvantages for the parent:

- Looking after the offspring increases the offspring's chance of surviving and passing on the parents' genes.
- However, looking after the offspring takes energy and makes the parents more at risk to predators.

Communication

EDEXCEL B3 ✓

Many animals live together in various types of social groups. This makes it necessary for them to communicate with each other. There are a number of ways that animals can do this:

- Sounds.
- Visual signals.
- Airbourne chemicals such as pheromones.

In mammals, such as chimpanzees, gorillas and humans, a lot of information is exchanged by **body language**. This includes gestures, body posture and facial expressions.

Make sure that you do not assume that animal behaviour means the same thing as human behaviour. That is called anthropomorphism. Chimps may look like they are smiling, but it is probably a sign of aggression.

The diagram shows some facial expressions in chimps, but the same expressions may mean something completely different to another type of animal. They are **species specific**.

Great ape behaviour

EDEXCEL B3 ✓

Humans are **great apes** along with gorillas, orangutans, chimpanzees and bonobos (pigmy chimps). Until recently little was known about the behaviour of great apes in the wild. People did not understand how they communicate and did not realise how intelligent they are. Two scientists studied great apes in the wild:

- Jane Goodall studied chimpanzees. She made two major discoveries. Firstly, like humans, chimpanzees enjoyed a mixed diet that included meat. Secondly, they could make simple tools, stripping leaves off branches to reach into termite nests. Until this time, only humans were thought to be clever enough to make tools.

- Dian Fossey studied mountain gorillas. She worked out many aspects of their communication and helped to protect them from poachers.

Do not mix up chimpanzees with monkeys – great apes are not monkeys. Apes are larger, spend more time upright, and depend more on their eyes than on their noses and do not have tails. They are more intelligent than monkeys.

PROGRESS CHECK

1. Why does a male peacock have such brightly coloured tail feathers?
2. What is unusual about albatross mating?
3. Write down one disadvantage of showing parental care.
4. What are pheromones?
5. What similarities between our hands and those of other great apes make tool use easier?
6. Dian Fossey is described as the first person to 'habituate herself with gorillas'. What does this mean?

1. So that it can compete for females and show that it is healthy and has good genes.
2. They pair for life.
3. It uses up food reserves and makes the parents more vulnerable to predators.
4. Airborne chemical messengers that are produced to attract a mate.
5. All apes have opposable fingers.
6. She lived with them and spent so much time with them they learned to ignore her. She was able to live with gorillas without them becoming alarmed by her presence.

Sample GCSE questions

1 The graph shows the brain mass and the body mass of a number of different mammals.

It includes modern man (*Homo sapiens*).

It also includes *Homo erectus*, an ancestor of modern man.

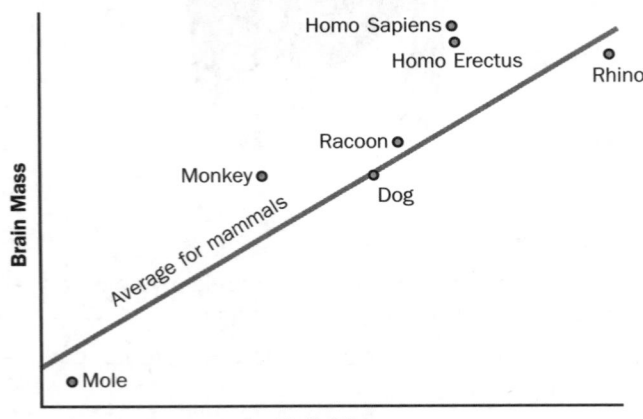

(a) How can scientists get the data they need about *Homo erectus* to plot this graph? **[2]**

By finding fossils and measuring the size of the skulls.

A good answer. Many candidates might just say 'from fossils' without explaining how.

(b) What pattern is shown by the graph? **[1]**

As the body mass increases then the brain size increases.

This is the general trend but you could say that there is some variation.

(c) What does the graph show about *Homo sapiens* and *Homo erectus* compared to other mammals? **[2]**

They are both close to the line of best fit so they are close fitting the pattern.

However they are slightly above the line which means that they have a slightly bigger brain for their size compared to other mammals.

(d) What evidence does the graph supply to help explain why *Homo erectus* is extinct today but *Homo sapiens* is not? **[2]**

Homo sapiens have a larger brain than Homo erectus had considering their size.

That means that they may have a greater ability to learn and so are more likely to survive.

Good answers. Drawing conclusions from graphs is a skill needed in many GCSE questions.

Exam practice questions

1 The diagram shows some steps in a baby as it learns language.

prefers voices to sounds	uses first words	uses simple sentences	uses grammar rules	ability to learn language stops
from birth	12 months	18 months	2 years	puberty

(a) Which part of the brain is involved in learning language?

.... **[1]**

(b) The diagram says that soon after it is born a baby prefers human voices to sounds.

Suggest why this is important for the baby.

....

.... **[1]**

(c) The diagram says that humans cannot learn language after puberty.

(i) Describe how the study of children that have grown up away from people (feral) has helped to support this idea.

....

.... **[2]**

(ii) Scientists have not performed experiments on children to prove this idea.

Suggest why.

....

.... **[1]**

2 As well as studying seagulls, the scientist Tinbergen also studied sticklebacks.

These are some of his observations.

The male stickleback marks out an area of sand on the bottom of the pond and digs a little hole.

Male sticklebacks will fight for the areas to build holes.

He gathers pieces of algae and piles them in the hole.

The male stickleback wiggles through the mount, leaving a tunnel.

Now the stickleback changes colour, becoming bluish white on the back and bright red on the underside.

Exam practice questions

The females, in the meantime, have become fat with hundreds of eggs.

When the male sights a female, he leads her to the nest.

The male prods near her tail and this causes her to lay eggs in the nest.

When she swims out, the male swims into the nest and fertilises the eggs.

He then guards the nest, fanning the eggs to keep them supplied with oxygen.

(a) Suggest why the male stickleback changes colour.

.... **[1]**

(b) The male sticklebacks fight for areas to build holes.

How can the population of sticklebacks gain from this behaviour?

....

.... **[2]**

(c) The male stickleback fans the eggs to supply them with oxygen.

What name is given to this type of behaviour?

.... **[1]**

(d) Tinbergen warned against anthropomorphism when looking at animal behaviour.

What does this mean?

.... **[1]**

3 Scientists have investigated behaviour in foxes.

They put a drug inside a dead rabbit.

The dead rabbit was then given to some foxes to eat.

The chemical made the foxes sick.

(a) The scientists found that the foxes are now frightened whenever they see a live rabbit.

What type of behaviour does this experiment demonstrate?

.... **[1]**

(b) This type of response increases the chance of foxes surviving.

Explain how.

....

.... **[1]**

Exam practice questions

4 A scientist was investigating memory.

(a) What happens in the brain when we learn something?

.... **[1]**

(b) The scientist showed a group of people fifteen pictures of objects.

The people had to remember the objects.

The people were shown the set of fifteen pictures twice.

The pictures were the same each time except that in the first showing one was a picture of a face.

This was then replaced by a picture of an object.

The graph shows how often the people correctly remembered each picture (% correct).

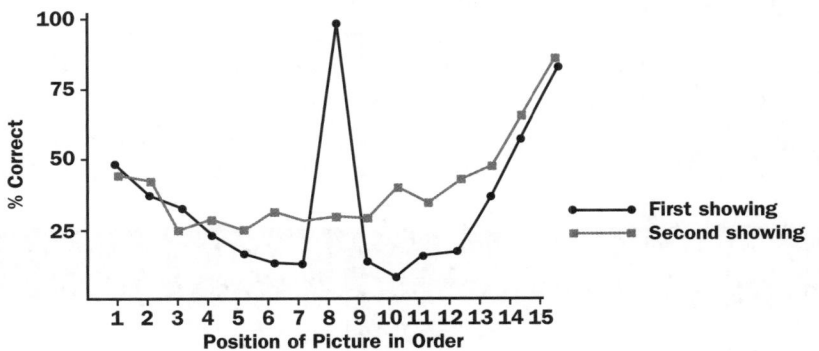

(i) The people remembered pictures 14 and 15 better than the other pictures.

Explain why.

....

.... **[1]**

(ii) Suggest which picture in the first showing was a face.

picture number .... **[1]**

(iii) Describe the difference between the results for the first and second showing and write about what happens in the brain that explains this difference.

The quality of written communication will be assessed in your answer to this question.

....

....

.... **[4]**

10 Microbes

The following topics are covered in this chapter:

- **The variety of microbes**
- **Putting microbes to use**
- **Microbes and genetic engineering**

10.1 The variety of microbes

LEARNING SUMMARY

After studying this section, you should be able to:

- describe the main types of microbes
- explain how microbes can cause disease
- describe the discoveries of Pasteur, Lister and Fleming with regard to microbes and disease
- describe the seasonal variations in plankton numbers.

Types of microbes

OCR B	B6	✓
CCEA	B2	✓
WJEC	B2	✓

KEY POINT

Microorganisms are organisms that are too small to be seen with the naked eye.

There is quite a variety of organisms that fall into this category:

- Fungi such as yeast.
- All bacteria.
- All viruses.
- Single celled members of the protoctista kingdom.

Fungi have certain features in common, but **yeast** is unusual because it is made up of single cells. Each cell has a nucleus, cytoplasm and cell wall. They reproduce asexually by growing a small **bud** on the side which breaks off and forms a new cell.

Bacteria are just a few microns (thousandths of a millimetre) in size. Their structure is shown on page 101. They have many different shapes such as spherical, rod, spiral and curved rods. Some bacteria can move using a **flagellum**. Bacteria all reproduce by a type of asexual reproduction called **binary fission**. Bacteria can survive on an enormous range of energy sources and can exploit a wide range of habitats as some take in food while others can photosynthesise.

> Think of a way of remembering the four types of microbes. You could use the first letters of their names, **F**, **B**, **V** and **P**.

Viruses are much smaller than bacteria and fungi and are usually considered to be non-living. They are made of a protein coat surrounding a strand of genetic material. Viruses can only reproduce in other living cells by injecting genetic material into the cell and using the cell to make the components of new viruses which then assemble into new virus particles.

Protoctista include organisms that used to be called protozoa and also single-celled algae such as plankton.

Microbes and disease

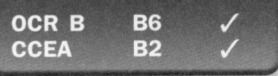

| OCR B | B6 | ✓ |
| CCEA | B2 | ✓ |

All the types of microbes include some organisms that are pathogens and cause disease. The ways that these microbes can be passed on is shown on page 35.

Examples of pathogens are:

- Bacteria such as *Salmonella* and *E. coli* that cause food poisoning.
- *Vibrio* bacteria that cause cholera.
- Fungi like *Trichophyton* that cause athlete's foot.
- Viruses that cause influenza and chicken pox.

For a pathogen to cause disease various steps usually happen:

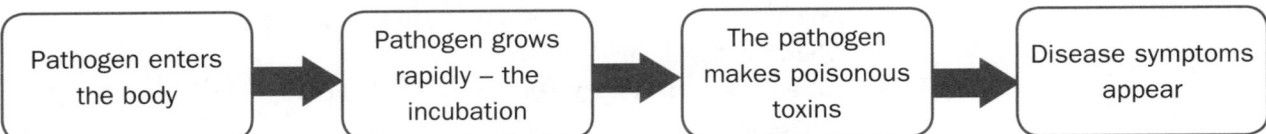

| Pathogen enters the body | → | Pathogen grows rapidly – the incubation | → | The pathogen makes poisonous toxins | → | Disease symptoms appear |

Diseases such as cholera can occur at any time, but after a natural disaster large numbers of people may become ill. This is because sewage and water systems may become contaminated and refrigerators may not work due to lack of electricity.

Modern medicine can treat many diseases. This is the result of discoveries made by many scientists. **Pasteur**, **Lister** and **Fleming** are three of these:

- Louis Pasteur studied a number of diseases such as rabies and anthrax. He was the first person to realise that diseases that can be passed on are caused by living organisms. This is called the **germ theory**.
- Joseph Lister worked as a doctor, operating on patients. He found that treating his instruments and washing his hands with a chemical called carbolic acid helped to stop his patients' wounds becoming infected. This was the first **antiseptic** to be used.
- Sir Alexander Fleming was growing bacteria on Petri dishes. He noticed that one of his dishes had a fungus growing on it. Around the fungus, called *Penicillium*, the bacteria had been killed. The fungus was producing a chemical called penicillin which was the first **antibiotic** to be discovered.

Plankton

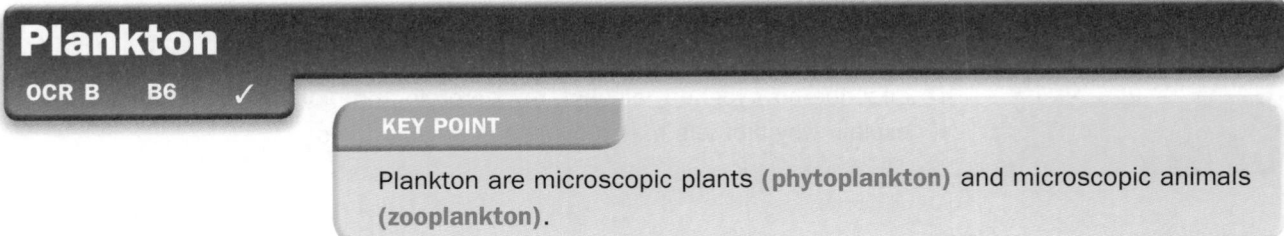

| OCR B | B6 | ✓ |

> **KEY POINT**
>
> Plankton are microscopic plants (**phytoplankton**) and microscopic animals (**zooplankton**).

Phytoplankton are capable of photosynthesis and are the main producers in ocean food chains and webs. The number of plankton living in lakes and the sea varies at different times of the year. These seasonal fluctuations can be explained by changes in light, temperature or minerals.

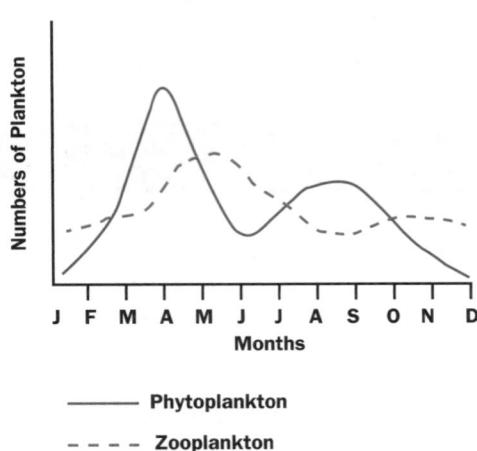

Seasonal Cycles in the North Atlantic

—————— Phytoplankton

– – – – Zooplankton

Be careful interpreting graphs such as this one. This is the North Atlantic so think about which months will be warmer and lighter. Graphs of more southern regions would be different.

PROGRESS CHECK

1. Why is reproduction in yeast often called 'budding'?
2. What is a flagellum?
3. Put a virus, a bacterium and a yeast cell in order of size with the largest first.
4. How did the antibiotic penicillin get its name?
5. Look at the graph of plankton numbers above.
   a) Suggest why the numbers of phytoplankton peak in April.
   b) Suggest why the numbers of zooplankton peak in May–June.

1. Because it produces small growths or buds that split off; a part of the yeast breaks off the parental cell.
2. A long whip-like projection from bacterial cells that allows them to move.
3. Yeast, bacterium, virus.
4. It is produced by the fungus *Penicillium*.
5. **a)** The light and temperature is increasing and there are plenty of minerals available.
   b) The numbers of phytoplankton have peaked and so there is plenty of food for them.

10.2 Putting microbes to use

LEARNING SUMMARY

After studying this section, you should be able to:

- explain the functions of the main parts of a fermenter
- describe how yoghurt, alcohol, mycoprotein and biofuels are made
- explain why biofuels may provide sustainable fuels.

Biotechnology

OCR A	B7	✓
OCR B	B6	✓
EDEXCEL	B3	✓
WJEC	B3	✓

Microorganisms have been used for hundreds of years for making foods such as bread and cheese. They are now being used more and more to produce new types of food and other useful products.

> **KEY POINT**
>
> The use of living organisms to make useful products is called **biotechnology**.

Microorganisms can be grown in large vessels called **fermenters**. The conditions are carefully controlled so that the microorganisms grow very fast. Like all organisms, they need food to obtain energy, but they can often be fed on waste from other processes.

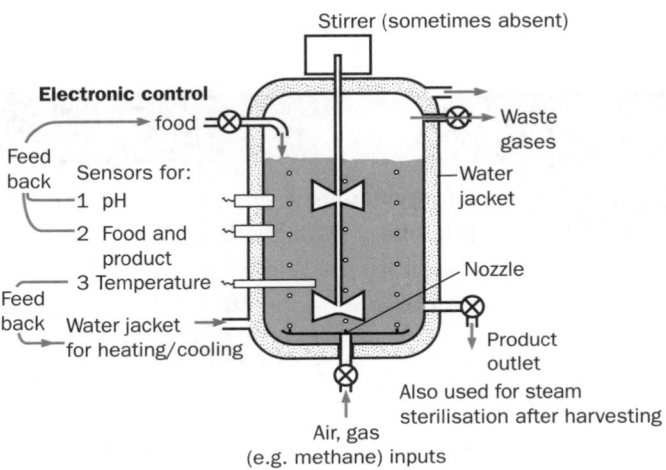

A fermenter.

Making yogurt

OCR B	B6	✓
EDEXCEL	B3	✓
WJEC	B3	✓

Energy source = milk.

Microbe = bacteria (*Streptococcus* and *Lactobacillus*).

Respiration = anaerobic.

Products = lactic acid.

> The lactic acid made by the bacteria makes the milk proteins clot to make the yoghurt thicker and gives it an acidic taste.

To make yoghurt the equipment is first sterilised and the milk is **pasteurised**. Pasteurisation involves heating milk to 78°C to kill harmful microbes. A culture of bacteria is added and the mixture is incubated at about 46°C for 4–5 hours. Before packaging, fruit, flavours or colours may be added.

Making alcohol

| OCR A | B4 | ✓ |
| OCR B | B6 | ✓ |

Energy source = grapes for wine or barley for beer.

Microbe = yeast.

Respiration = anaerobic.

Products = alcohol and carbon dioxide.

Making alcohol relies on yeast fermenting sugar. This sugar can come from different sources. The equation for fermentation is:

glucose → ethanol (alcohol) + carbon dioxide
$$C_6H_{12}O_6 \rightarrow C_2H_5OH + 2CO_2$$

In beer making, barley is allowed to germinate so that the starch is turned into sugar. Hops are added to give the beer flavour. After the yeast has fermented, the beer is drawn off the yeast. It may then be pasteurised to kill any microbes before it is bottled.

The concentration of alcohol made by fermentation is limited. This is because when the alcohol reaches a certain concentration it will kill the yeast. Spirits are made by **distillation**. This concentrates the alcohol. The type of spirit that is produced depends on the sugar source:

- Rum from sugar cane.
- Whisky from malted barley.
- Vodka from potatoes.

Making mycoprotein

AQA	B3	✓
EDEXCEL	B3	✓
WJEC	B3	✓

Energy source = any starch or sugar source such as potato waste.
Microbe = a fungus (*Fusarium*).
Respiration = aerobic.
Products = mycoprotein.

Mycoprotein is often used as a meat substitute as it has a high protein content. It does have a number of advantages over meat:

- The fungus grows very quickly.
- It has a high fibre content.
- It is low in fats.
- It can be grown on waste substances.

Making biofuels

AQA	B3	✓
OCR A	B4	✓
OCR B	B6	✓
EDEXCEL	B3	✓
WJEC	B3	✓

KEY POINT

Biofuels are fuels that have been made from living material that has been produced in a sustainable way.

Biogas is produced in fermenters called **digesters**. Waste materials such as sewage or plant products are put in the tank. A mixture of different bacteria use these substances for anaerobic respiration. They produce biogas which is made up of:

- largely methane
- some carbon dioxide
- small amounts of hydrogen, nitrogen and hydrogen sulphide.

The biogas can then be burned to make electricity, produce hot water or power motor vehicles.

Some countries can grow large amounts of sugar cane or maize. These crops can be used to produce sugar that can be fermented by yeast. This produces **ethanol**. The ethanol can be concentrated by distillation and added to petrol.

petrol (about 85%) + ethanol (about 15%) → gasohol

Cars can be converted to run on the gasohol and so less petrol is used.

Biofuels and sustainability

AQA	B3	✓
OCR A	B4	✓
OCR B	B6	✓
EDEXCEL	B3	✓

Fossil fuels such as coal and oil have been produced from living material over millions of years. They are not sustainable as we are burning them faster than they are being produced. This means that carbon dioxide is being added to the air.

Microbes can be used to produce biofuels such as **biogas** and **ethanol**. These fuels are sustainable because they release carbon dioxide to the air at the same rate that it is being absorbed by photosynthesis. However, people are worried that in some areas tropical forests might be being destroyed to grow biofuel crops.

PROGRESS CHECK

1. When making yoghurt, the equipment is sterilised first. Why is this?
2. After pasteurisation when making yoghurt, the milk is cooled to 46°C before the culture is added. Why is this?
3. Why can vegetarians use mycoprotein as a meat substitute?
4. Fermentation can only make wine containing up to about 15% alcohol. Why is this?
5. 'Energy forests' grow trees that are regularly harvested and burned as fuel. Why is this sustainable?
6. Suggest why people are so worried about the destruction of tropical forests.

6. It is a habitat containing many rare species/it may contribute to global warming.
5. The trees are only burned at the same rate that they grow so there is no net carbon dioxide release.
4. Because the alcohol kills the yeast when it reaches about 15%.
3. Because it is made from fungi not animals and is high in protein.
2. High temperatures would kill the culture.
1. So that any pathogens are killed.

10.3 Microbes and genetic engineering

LEARNING SUMMARY

After studying this section, you should be able to:

- describe some uses of genetic engineering
- explain the roles of the enzymes used in genetic engineering
- explain how insulin and chymosin are produced
- describe how plants can be modified for herbicide and insect resistance
- describe how enzymes can be used to produce genetic fingerprints.

Uses of genetic engineering

AQA	B1	✓
OCR A	B7	✓
OCR B	B6	✓
EDEXCEL	B2, B3	✓
CCEA	B2	✓

It is now possible to make microbes produce different proteins by changing their DNA. They are then called **genetically modified (GM)**. If the gene has come from a different species then the GM organism is called **transgenic**. The main steps in making a GM organism involve:

- Finding and removing the necessary gene.
- Putting the gene into a vector that will carry it into the new cell.
- Testing the cells to see if they have taken up the gene.
- Then allowing the new cells to reproduce and produce the new protein.

GM organisms can have lots of possible uses:

GM microbes	GM plants	GM animals	GM humans
Making chymosin	Resistance to weedkillers	Faster growth	Gene therapy
Making insulin	Increasing yield	Producing useful proteins in milk	
Making human growth hormone	Produce insecticide		
	Resistance to disease		

Some of the arguments for and against genetic engineering are on page 64.

Using enzymes in genetic engineering

OCR A	B7	✓
OCR B	B6	✓
EDEXCEL	B3	✓
CCEA	B2	✓

A number of steps in genetic engineering involve the use of enzymes. An example of this is the modification of rice so that it produces vitamin A. Two different enzymes are used:

- **Restriction enzymes** cut the DNA in specific places.
- **DNA ligase** joins DNA together.

How to produce genetically engineered rice.

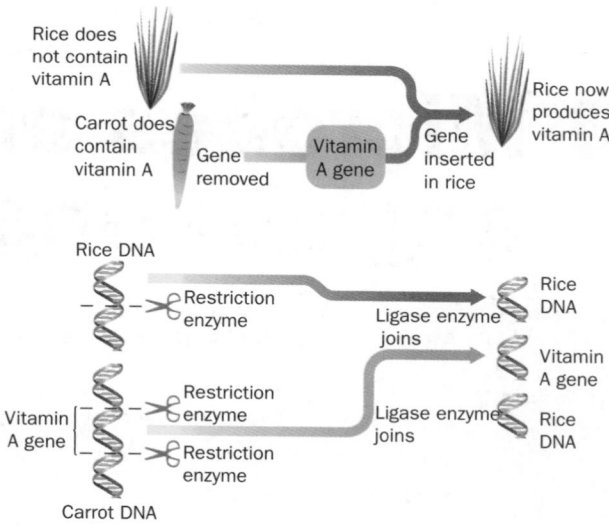

The restriction enzymes do not make a straight cut in the DNA, but make a staggered cut. This produces exposed unpaired bases. They are called **sticky ends** because any piece of DNA with complementary unpaired bases will stick to it.

Insulin and chymosin production

OCR A	B7	✓
OCR B	B6	✓
EDEXCEL	B2, B3	✓
CCEA	B2	✓

> Be careful when answering questions about insulin. Remember it is the insulin gene that is put into bacteria not the insulin molecule. This is a common mistake.

Insulin is needed to treat people who have diabetes. This is explained on page 13. Until 1982, the insulin used was extracted from cows or pigs. This was difficult to extract and worked slower than human insulin and the body often rejected it. Today, the human gene for insulin is placed into bacteria and produces insulin that is identical to human insulin. This works faster and can be produced in large quantities and will not be rejected as it is identical to human insulin.

Chymosin is an enzyme that has been used in cheese making for many years. It is produced in the stomach of young mammals and makes milk proteins clot. To make cheese it has been necessary to kill animals and extract the chymosin from their stomach. Today, most chymosin is made by GM microbes, usually fungi. They have had the gene for chymosin inserted. This means that large amounts of chymosin can be made quickly and the cheese can be eaten by vegetarians.

Herbicide and insect resistance

AQA	B1	✓
OCR A	B7	✓
OCR B	B6	✓
EDEXCEL	B2, B3	✓
WJEC	B1	✓

It is possible to use GM bacteria to produce GM plants. The bacteria are used to inject useful genes into plants. The type of bacteria used (*Agrobacterium tumefaciens*) injects its genetic material into plant cells producing a growth of plant cells called a gall. The cells of this gall can be grown into a complete GM plant.

> Remember the plasmid or the *Agrobacterium* are often called vectors because they carry the gene into the plant. This is a little like vectors carrying pathogens.

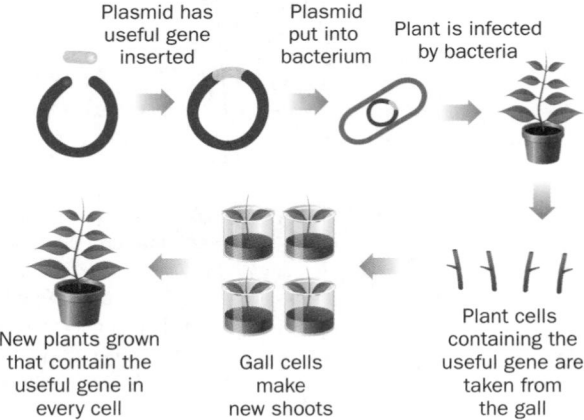

Plasmid has useful gene inserted → Plasmid put into bacterium → Plant is infected by bacteria → Plant cells containing the useful gene are taken from the gall → Gall cells make new shoots → New plants grown that contain the useful gene in every cell

Several different genes can be inserted:

- A gene may make the plant resistant to weedkillers (herbicides). This means the farmer can spray his fields to kill weeds without killing the crop.
- A gene from another type of bacteria (*Bacillus thuringiensis*) can be inserted. This makes the crop plant make its own insecticide.

Gene probes

AQA	B2	✓
OCR A	B7	✓
OCR B	B6	✓

Using the enzymes involved in genetic engineering it is possible to identify specific lengths of DNA from a person:

- Step 1 – the regions of the DNA are isolated and cut up using enzymes.
- Step 2 – the DNA fragments are put on a gel and separated using an electric current.
- Step 3 – the fragments are treated with a radioactive or fluorescent probe.

This technique could be used in gene testing to see if a person has a specific harmful allele.

It can also be used to make a **genetic fingerprint**.

> **KEY POINT**
>
> A genetic fingerprint is a pattern of DNA that can be used to identify an individual.

Scientists have discovered that our DNA contains regions that do not code for proteins. This is often called junk DNA although this is changing as biologists are realising it is important. In this DNA are regions with repeating sequences that can be used to identify a person.

PROGRESS CHECK

1. What is a transgenic organism?
2. Why should farmers want to make their crops produce insecticide?
3. Write down one advantage of using human insulin rather than pig insulin to treat diabetes.
4. Why can cheese made with GM chymosin be eaten by vegetarians?
5. Farmers grow crops that have a gene for resistance to weed killer inserted using *Agrobacterium*. Why do they want to grow these crops?
6. Why is a radioactive or fluorescent probe used in genetic testing?

1. An organism that contains genes from another species.
2. So that any insect pests will be killed if they try to eat the plants, so that they do not have to waste time and effort spraying the crops as the pesticide is already there.
3. Works better/has fewer side effects/some religious faiths (e.g. Jewish) would be able to have the insulin.
4. It does not involve killing animals to extract the chymosin from their stomachs.
5. They can then spray their crops with weed killer and only the weeds will die, therefore reducing competition.
6. So that the sections of DNA can be seen/photographed; to show where the gene being tested for is.

Sample GCSE questions

1 Read the passage about gasohol and answer the questions that follow.

> In 1972 the price of oil went up dramatically and this made many countries look for alternative fuels.
>
> A programme was set up in Brazil to produce ethanol by fermentation of sugar cane followed by distillation.
>
> The ethanol can be used on its own or in a mixture called gasohol.
>
> Now scientists are working on a genetically modified yeast that can digest and ferment starch.
>
> Other countries are starting to grow large crops of plants to produce palm oil. This can be used as a fuel.

(a) What is ethanol mixed with to produce gasohol? **[1]**

Petrol

(b) Apart from cost, explain one advantage of using gasohol rather than pure fossil fuels. **[2]**

Fossil fuels such as oil are running out so mixing them with ethanol means that they will last longer.

> This is a reasonable point but the key reason is that burning ethanol only gives out the same amount of carbon dioxide that the sugar cane took in. It is carbon neutral.

(c) Why is producing gasohol more feasible in hot countries such as Brazil, rather than in Britain? **[1]**

It is too cold in Britain to grow sugar cane.

(d) How might the genetically engineered yeast allow biofuels to be used by more people? **[2]**

Countries that cannot grow sugar cane can grow many other crops that contain starch such as potatoes.

This can be digested and fermented by the yeast.

(e) Describe the problems that can be caused by some countries growing large plantations producing crops such as palm oil. **[3]**

In some countries large areas of natural vegetation such as rainforests are being cut down to grow biofuel crops.

This is causing animals and plants to be threatened by extinction.

It may also cause changes to the climate.

> So although biofuels are grown so that we do burn less fossil fuels and give out less carbon dioxide, in some areas the burning of rainforests is cancelling this out because less carbon dioxide is being taken in.

Exam practice questions

1 The diagram shows the main stages in yoghurt making.

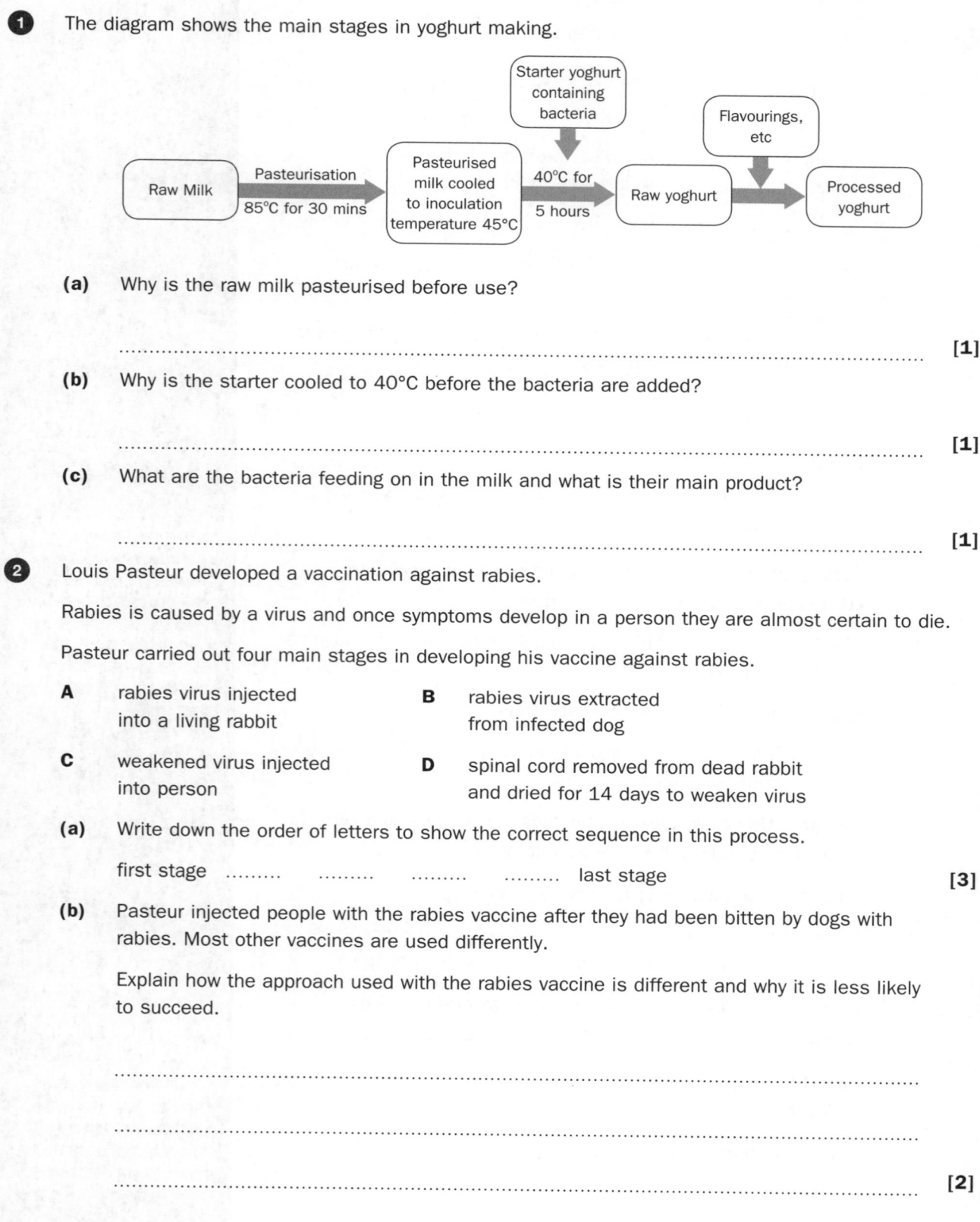

(a) Why is the raw milk pasteurised before use?

.... **[1]**

(b) Why is the starter cooled to 40°C before the bacteria are added?

.... **[1]**

(c) What are the bacteria feeding on in the milk and what is their main product?

.... **[1]**

2 Louis Pasteur developed a vaccination against rabies.

Rabies is caused by a virus and once symptoms develop in a person they are almost certain to die.

Pasteur carried out four main stages in developing his vaccine against rabies.

A rabies virus injected into a living rabbit

B rabies virus extracted from infected dog

C weakened virus injected into person

D spinal cord removed from dead rabbit and dried for 14 days to weaken virus

(a) Write down the order of letters to show the correct sequence in this process.

first stage .... last stage **[3]**

(b) Pasteur injected people with the rabies vaccine after they had been bitten by dogs with rabies. Most other vaccines are used differently.

Explain how the approach used with the rabies vaccine is different and why it is less likely to succeed.

....

....

.... **[2]**

Exam practice questions

3 Chymosin has traditionally been extracted from cows.

It can be produced by bacteria using genetic engineering.

The diagrams show four stages of this process.

The diagrams are not in the correct order.

A

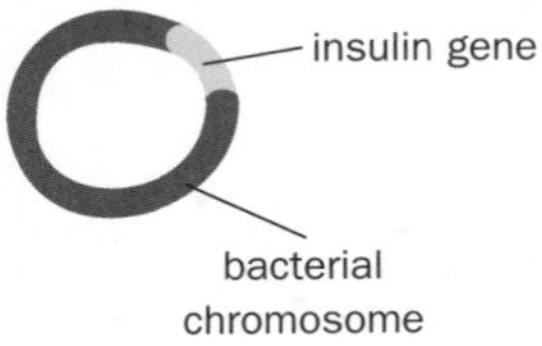

insulin gene

bacterial chromosome

B

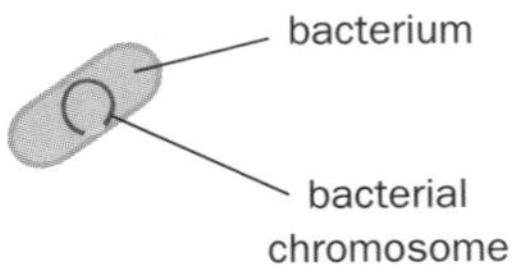

bacterium

bacterial chromosome

C

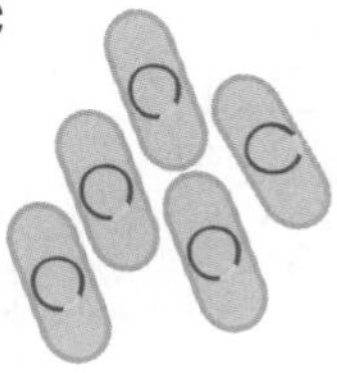

D

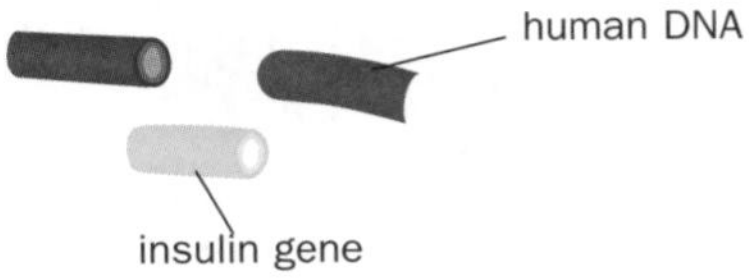

human DNA

insulin gene

(a) What is chymosin used for?

.... [1]

(b) Put the stages in order and describe what is happening at each stage.

Stage	Letter	Description of stage
1		
2		
3		
4		

[7]

(c) Write down **two** advantages of producing chymosin by genetically engineered bacteria rather than extracting it from cows.

....

.... [2]

Answers

Note: For questions involving QWC, marks will be awarded if:
- All information in answer is relevant, clear, organised and presented in a structured and coherent format.
- Specialist terms are used appropriately.
- There are few, if any, errors in grammar, punctuation and spelling.

Chapter 1

1. **(a)** Vitamin C.
   (b) She needs $8 \times 60 = 480$ mg.
   She only takes in 450 mg and so may suffer from anaemia.
   (c) Babies need more protein per kg of body mass.
   This is because they are growing more.
   If the 30 year-old is pregnant then she needs more because she needs to supply protein to the baby so it can grow.
   (d) Iron is needed to make haemoglobin.
   The girl will need to replace haemoglobin lost during periods.
   (e) (i) $60 \times 0.6 = 36$ g.
   $\frac{9}{36} \times 100 = 25\%$.
   (ii) This will not cause an increase in blood pressure; and so will not increase the risk of a stroke/heart disease.

2. **(a) (i)**

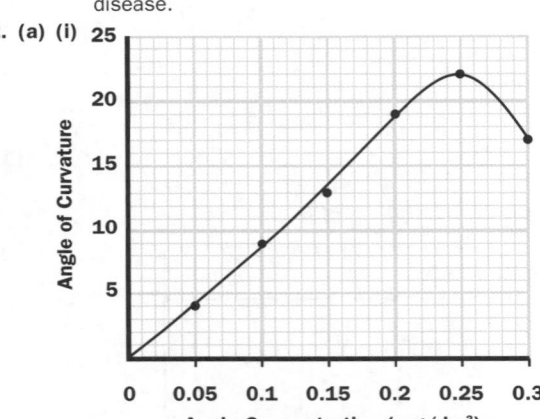

   (Marks awarded for suitable scale on *y* axis, Points correctly plotted; best curve drawn.)
   (ii) Increased auxin concentration causes more curvature.
   Levels off/decreases at high concentrations.
   Figures quoted from graph.
   (b) (i) It causes cells to elongate.
   More on the side where the block is placed.
   (ii) Light; negative geotropism/gravitropism.
   (iii) Leaves can trap more sunlight/more photosynthesis.
   So more food production/more growth.
   (c) Rooting of cuttings/weedkillers/seedless fruits/control of fruit ripening.

3. **(a)**

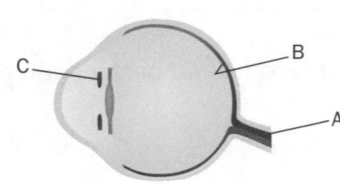

   (b) The lens is too powerful (does not refract the light enough).
   The light is focussed behind the retina.

4.

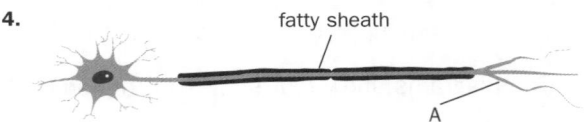

   (c) Touch, smell and vision use sensory neurones.
   These are not affected.
   Breathing involves motor neurones going to the intercostal muscles.

Chapter 2

1. **(a)**

The chemicals made by the bacteria that are making Vijay feel ill		Antibiotics
An injection that could have stopped Vijay getting the disease		Vaccination
The medicine given to Vijay to kill the bacteria		White blood cells
The cells produced by Vijay's body to kill the bacteria		Toxins

   (b) Destroyed by hydrochloric acid.
   (c) Sebum on skin/enzyme in tears/mucus in the breathing system/skin.
   (d) Proved that it worked/showed that it was not harmful.

2. **(a)** Heavy smokers have a higher annual death rate.
   Non-smokers do not seem to get lung cancer.
   Heavy smokers have a higher risk of getting lung cancer than all patients.
   (b) Two from: Doll made more observations.
   Better communication.
   One example of communication method.
   More awareness.

3. **(a) (i)** It often contains a weakened form of the microbe.
   The person gets a mild form of the disease.
   (ii) Two from: worried about the side effects.
   Three sets of side effects at once may damage the child.
   Reference to autism.
   (b) Two from: one vaccine protects children more quickly and effectively than one, as child is protected from start.
   Cheaper.
   Takes less time.
   May not come back for other injections.
   (c) Contains three different weakened pathogens/antigens.
   Each one stimulates the production of a different antibody.
   These antibodies/memory cells will stay in the blood.

4. **(a)** Overall trend is an increase in deaths from 1928 to 1955.
   Some fluctuations.
   (b) There is a drop in deaths for a few years after 1939.
   Fat in the diet is a risk factor for heart disease.
   Blocks the coronary arteries.
   (c) Probably did more exercise/less stress.

Answers

Chapter 3

1. **(a)** Asexual reproduction.
   (b) Two from: quicker to produce a full grown plant.
   Only need one parent.
   Plants will be identical to parent plant.
   (c) Tissue culture is an artificial process/only requires a small number of cells.
2. **(a)** Changes to DNA/genes.
   (b) Wear sunglasses/use sunscreen/do not go out in bright sunshine.
   (c) It does not express itself if the dominant allele is present.
   Only people who are homozygous for the allele are albinos.
   (d) Genotypes of the two parents: Aa x Aa.
   Gametes: A or a x A or a.
   Offspring genotypes AA, Aa, Aa, aa
   Indication that aa is albino.
3. **(a)** Peter's gametes: X and Y.
   Correct offspring: XX, XX, XY, XY.
   (b) (i) All countries.
   All ratios after the age of 65 show less than one man to each woman.
   (ii) India.
   The ratio of boys to girls at birth is highest/above one to one.
   (iii) Imbalance in the sex ratio in the country leading to possible drop in population.
4. **(a)** See predators approaching/see prey.
   (b) (see QWC guidance on page 194) All apes were born with slightly different bones/mutations.
   The apes that could stand the best could see further and were more likely to survive.
   They could reproduce and pass on their genes.
   Over many generations the population gradually became more upright.
   (c) Unlikely that tool use was the reason for man being upright.
   Because fossil was upright but could not have used tools as the brain was too small.

Chapter 4

1. **(a)** A = nitrates; B = nitrogen gas; C = ammonium compounds.
   (b) (i) Fungi.
   (ii) Any three from: a reasonable temperature.
   A suitable pH.
   Oxygen.
   Moisture.
   (c) (see QWC guidance on page 194) Lives in root nodules.
   Mutualistic relationship.
   The plant gains nitrogen compounds from the bacteria.
   The bacteria gains some carbohydrates from the plant.
2. **(a) (i)**

Kingdom	Animal
Phylum	Vertebrate
Class	Mammal
Order	**Cetartiodactyla**
Family	**Camelidae**
Genus	Camelus
Species	dromedaries

   (ii) To avoid confusion.
   Common names vary in different places.
   (b) (i) Stop it sinking in the sand.
   (ii) Two from: allows them to go without food for some time as it is scarce.
   Can be respired to produce water.
   Being on the top of the body insulates it from the Sun.
3. **(a)** $3050 - (2000 + 925) = 125$ kJ.
   (b) $\frac{125}{3025} \times 100 = 4.13\%$.
   (c) Food has to go through two energy conversions/trophic levels if we eat meat.
   More energy is lost.

Chapter 5

1. **(a) (i)** $6O_2$; $6CO_2$
   (ii) Three from: to supply the body with more oxygen.
   Because respiration rate has increased in the muscles.
   To prevent anaerobic respiration.
   To remove the extra carbon dioxide.
   (b) glucose $\rightarrow$ lactic acid + energy
   (c) There is still excess carbon dioxide to be removed.
   Lactic acid has to be broken down.
   Payback the oxygen debt.
2. **(a)** The rate of reaction steadily increases as the temperature increases.
   Peaks at about 43°C.
   Above that the rate drops rapidly.
   (b) Three from: up to the optimum any increase in temperature increases the kinetic energy.
   Molecules collide more/collide with more energy.
   Above optimum the enzymes start to denature.
   Substrate cannot fit into active site.
   (c) Enzyme A is from the human because its optimum is closest to 37°C, body temperature.
   Enzyme B is from the bacterium as it can work at high temperatures.
3. **(a)** Three from: the two strands of DNA unwind and unzip.
   Each strand attracts complementary bases.
   The bases are joined together and each molecule winds up.
   Each new molecule has one old and one new strand.

   (b)

Feature	Mitosis	Meiosis
Number of cells made from one cell	**Two**	Four
Uses of cells that are made	Growth /repair / asexual reproduction	**Sex cells**
Number of chromosomes in the cells made	**Same number as the parent cells**	Half the number of the parent cell

   (c) In the first division the whole chromosomes from each pair move apart.
   In the second division the copies of each chromosome move apart.

Answers

Chapter 6

1. (a) (i) This is the place that organisms live, i.e. the pond.
  (ii) All the organisms of one species living in the pond.
  (iii) All the living organisms in the pond plus all the non-living components e.g. water and soil.
  (b) (i) A net.
  (ii) Area is $3.142 \times 7.5^2 = 176.74$ m².
  Population = $5 \times 176.74 = 884$ snails.
  (iii) $30 \times \frac{29}{2} = 435$ (1 mark for calculation and 1 mark for answer)
  (iv) The second method because it samples in 5 areas. One area might not be representative.
2. (a) (i) Tree gained 78 kg but soil only lost 1 kg.
  All the mass could not have come from the soil.
  (ii) Water is needed for photosynthesis.
  But so is carbon dioxide.
  And minerals are also needed from the soil.
  (b) Three from: he would use telephones; letters; scientific journals; lectures.
  van Helmont would not have used the telephone.
3. (a) (i) It may leave toxic residues/many kill pollinating insects/may kill predators of red spiders.
  (ii) Spider population is developing resistance.
  (b) (i) Biological control.
  (ii) Must make sure that the introduced organism; does not become a pest.
  (c) Hydroponics.
  Easy to provide the plant with minerals.
  Can include insecticide in the water.

Chapter 7

1. (a) It loses mass at a steady rate.
  Because it is losing water.
  (b) Lose less mass than the unpainted leaf because nail varnish prevents water loss.
  Less mass lost when bottom surface is painted because stomata are found there.
  Little difference if top surface painted as already a waxy cuticle present.
  (c) More wind movement causes more water loss.
  It blows away the water that has diffused out.
2. (a) Blood is pumped by the heart.
  It does carry oxygen to the tissues.
  However it returns in different vessels called veins.
  (b) The veins are carrying blood back to the heart from the arm and so blood gets trapped.
  (c) (i) Valves.
  (ii) The blood is under higher pressure in arteries so would not flow backwards.
  (d) Capillaries.
  His microscope was not powerful enough.
3. (a) (i) Absorption of digested food.
  (ii) Two from: large surface area/long.
  Villi/folds/microvilli.
  Rich blood supply.
  (b) (see QWC guidance on page 194) Emulsifies fat.
  Increases the surface area of the droplets.
  Allows lipase to work faster.
  Neutralises acid from stomach.

Chapter 8

1. (a)

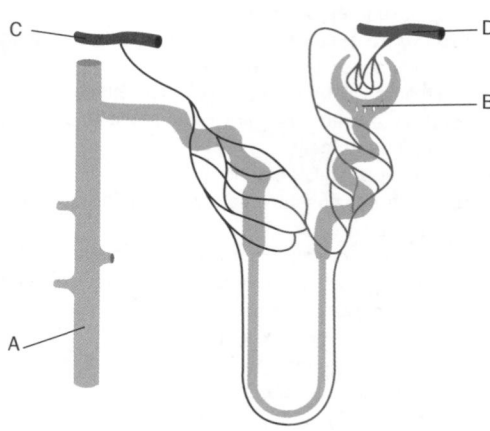

  (b) (i) Protein is too large to be filtered out of the blood.
  (ii) Glucose is filtered.
  But is then reabsorbed.
  (iii) Some of the glucose has been used for respiration; by the kidney.
  (c) The kidney filters this out of the blood.
  It is excreted in the urine.
2. (a) Prevents clots and bubbles being returned to the body.
  Which could then block blood vessels.
  (b) Waste substances from the blood diffuse out.
  Through the partially permeable membrane into the dialysis fluid.
  (c) It must contain the same concentration of glucose/amino acids/salts as plasma.
  Because these substances must not diffuse out of the blood.
3. (a) (i) Muscle on the right labelled biceps.
  (ii) The arm flexes/bends.
  (iii) Via tendons.
  (iv) They work against each other.
  When one contracts the other relaxes.
  (b) (i) Synovial/hinge joint.
  (ii) Reduce friction.
  Absorb shock.
  (c) Strong.
  Long lasting.
  (d) (i) More 56 to 70 year olds than younger people as joints wear with age.
  Fewer older people alive/may not risk operation.
  (ii) $\frac{16}{36} \times 100 = 44.4\%$ (1 mark for calculation and 1 mark for answer)
  (iii) Generally fitter/tissues repair faster when younger.

Chapter 9

1. (a) Cerebral cortex/cerebrum.
  (b) So that it can recognise/respond to its mother.
  (c) (i) Children have been discovered who have not been exposed to language up to puberty.
  They have then not been able to learn language successfully.
  (ii) Not ethical to separate children from people.

Answers

2. **(a)** To attract the females.
   (b) Strongest males will mate; and pass on their successful genes.
   (c) Parental care.
   (d) Assuming that animals think and behave in the same way as humans do.
3. **(a)** Conditioning.
   (b) They learn to avoid harmful stimuli/food in this way.
4. **(a)** New neural connections/pathways/synapses are formed.
   (b) (i) They were the last few that were shown.
   (ii) 8.
   (iii) (see QWC guidance on page 194) More correct answers for the second showing.
   This is because neural pathways are reinforced.
   This improves learning.

Chapter 10

1. **(a)** To kill harmful microbes.
   (b) So that the bacteria are not killed.
   (c) Lactose.
   Lactic acid.
2. **(a)** B A D C.
   (b) Vaccines usually used before microbe is encountered. Therefore can build up antibodies and so kill the microbe before it can cause harm.
3. **(a)** Cheese making.
   (b) D A B C.
   D The chymosin gene is cut out of cow DNA.
   A The gene is inserted into a bacterial plasmid.
   B The plasmid is inserted into a bacterium.
   C The bacteria reproduces and makes cow chymosin.
   (c) Two from: Can be eaten by vegetarians.
   Can produce large amounts.
   Bacteria can be grown on waste.
   Can be produced at any time of the year.

Notes